SpringerBriefs in Molecular Science

Chemistry of Foods

Series editor

Salvatore Parisi, Industrial Consultant, Palermo, Italy

The series Springer Briefs in Molecular Science: Chemistry of Foods presents compact topical volumes in the area of food chemistry. The series has a clear focus on the chemistry and chemical aspects of foods, topics such as the physics or biology of foods are not part of its scope. The Briefs volumes in the series aim at presenting chemical background information or an introduction and clear-cut overview on the chemistry related to specific topics in this area. Typical topics thus include: - Compound classes in foods – their chemistry and properties with respect to the foods (e.g. sugars, proteins, fats, minerals, …) - Contaminants and additives in foods – their chemistry and chemical transformations - Chemical analysis and monitoring of foods - Chemical transformations in foods, evolution and alterations of chemicals in foods, interactions between food and its packaging materials, chemical aspects of the food production processes - Chemistry and the food industry – from safety protocols to modern food production The treated subjects will particularly appeal to professionals and researchers concerned with food chemistry. Many volume topics address professionals and current problems in the food industry, but will also be interesting for readers generally concerned with the chemistry of foods. With the unique format and character of Springer Briefs (50 to 125 pages), the volumes are compact and easily digestible. Briefs allow authors to present their ideas and readers to absorb them with minimal time investment. Briefs will be published as part of Springer's eBook collection, with millions of users worldwide. In addition, Briefs will be available for individual print and electronic purchase. Briefs are characterized by fast, global electronic dissemination, standard publishing contracts, easy-to-use manuscript preparation and formatting guidelines, and expedited production schedules. Both solicited and unsolicited manuscripts focusing on food chemistry are considered for publication in this series.

More information about this series at http://www.springer.com/series/11853

Caterina Barone · Marcella Barbera
Michele Barone · Salvatore Parisi
Aleardo Zaccheo

Chemical Evolution of Nitrogen-based Compounds in Mozzarella Cheeses

Caterina Barone
COIF Association
Palermo
Italy

Marcella Barbera
Agency for the Environmental Protection
(ARPA) Sicilia
Ragusa
Italy

Michele Barone
COIF Association
Palermo
Italy

Salvatore Parisi
Industrial Consultant
Palermo
Italy

Aleardo Zaccheo
Bioethica Food Safety Engineering Sagl
Lugano-Pregassona
Switzerland

ISSN 2191-5407 ISSN 2191-5415 (electronic)
SpringerBriefs in Molecular Science
ISSN 2199-689X ISSN 2199-7209 (electronic)
Chemistry of Foods
ISBN 978-3-319-65737-0 ISBN 978-3-319-65739-4 (eBook)
DOI 10.1007/978-3-319-65739-4

Library of Congress Control Number: 2017950003

© The Author(s) 2018
This work is subject to copyright. All rights are reserved by the Publisher, whether the whole or part of the material is concerned, specifically the rights of translation, reprinting, reuse of illustrations, recitation, broadcasting, reproduction on microfilms or in any other physical way, and transmission or information storage and retrieval, electronic adaptation, computer software, or by similar or dissimilar methodology now known or hereafter developed.
The use of general descriptive names, registered names, trademarks, service marks, etc. in this publication does not imply, even in the absence of a specific statement, that such names are exempt from the relevant protective laws and regulations and therefore free for general use.
The publisher, the authors and the editors are safe to assume that the advice and information in this book are believed to be true and accurate at the date of publication. Neither the publisher nor the authors or the editors give a warranty, express or implied, with respect to the material contained herein or for any errors or omissions that may have been made. The publisher remains neutral with regard to jurisdictional claims in published maps and institutional affiliations.

Printed on acid-free paper

This Springer imprint is published by Springer Nature
The registered company is Springer International Publishing AG
The registered company address is: Gewerbestrasse 11, 6330 Cham, Switzerland

Contents

Chapter 1
Biogenic Amines in Cheeses: Types and Typical Amounts

Abstract This chapter evaluates the consequences of protein modifications in cheeses with specific relation to the production of biogenic amines and related influence on food quality and safety. As certain biogenic amines display a toxic potential to humans, considerable research has been carried out in recent years to evaluate their present in fermented foods such as cheeses. The presence of amines is influenced by different factors such as cheese variety, seasoning and microflora. With specific relation to cheeses, the main biogenic amines analytically detected in cheeses are histamine, tyramine, putrescine, cadaverine and 2-phenylethylamine. These biogenic amines are discussed from the chemical viewpoint; also, technological aspects of cheesemaking productions are analysed in connection with biogenic amines production. Consequently, adequate mitigation strategies are needed because safety is a basic requirement in food productions; in addition, the current legislation defining biogenic amines and related tolerances in fermented foodstuffs does not appear sufficient.

Keywords Amino acid decarboxylase · Cheesemaking · Decarboxylating activity · Intoxication · Mitigation strategy · Ripening time · Risk assessment

Abbreviations

Agmatine	AGM
Amino acid	AA
Amino acid decarboxylase	AAD
Biogenic amine	BA
Cadaverine	CAD
Carbon dioxide	CO_2
Diamine oxidase	DAO
European Food Safety Authority	EFSA
Food and Agriculture Organization of the United Nations	FAO
Food and Drug Administration	FDA

C. Barone, M. Barbera, M. Barone, S. Parisi, A. Zaccheo, *Chemical Evolution of Nitrogen-based Compounds in Mozzarella Cheeses,* SpringerBriefs in Chemistry of Foods.

© The Author(s) 2018

C. Barone et al., *Chemical Evolution of Nitrogen-Compounds in Mozzarella Cheeses*, SpringerBriefs in Molecular Science,
DOI 10.1007/978-3-319-65739-4_1

High-performance liquid chromatography	HPLC
Histamine	HIS
Lactic acid bacteria	LAB
Monoamine oxidase	MAO
Non-starter lactic acid bacteria	NS-LAB
Oxygen:	O_2
2-phenylethylamine	PHE
Putrescine	PUT
Serotonin	SER
Sodium chloride	NaCl
Spermidine	SPD
Spermine	SPM
Tryptamine	TRP
Tyramine	TYR

1.1 Biogenic Amines in Foods. Chemical Structures and Biosynthetic Pathways

Biogenic amines (BA) constitute biological active, non-volatile, low-molecular weight, organic bases. These compounds can be produced and degraded as the result of normal metabolic activities in humans, animals, plants and microorganisms. Bioactive amines are also present in a variety of food products like chocolate, meat, fish and fermented foods (beer, wine and dairy products) [1]. The main important compounds in this category are: histamine (HIS), tyramine (TYR), serotonin (SER), tryptamine (TRP), putrescine (PUT), cadaverine (CAD), 2-phenylethylamine (PHE), agmatine (AGM), spermidine (SPD) and spermine (SPM). The first six compounds are shown in Fig. 1.1.

According to their chemical structure, BA can be classified as aromatic (histamine, tiramine, serotonin, 2-phenylethylamine and tryptamine) on the one hand, and aliphatic compounds on the other side (putrescine, cadaverine, agmatine, spermidine and spermine). In addition, BA can be discriminated depending on the number of amino groups: the proposed classification concerns monoamines (2-phenylethylamine and tyramine), diamines (histamine, serotonin, putrescine and cadaverine) and polyamines (spermine, spermidine and agmantine) [2]. These amines may arise through microbial processes by decarboxylation of free precursor amino acids (AA) or by amination and transamination of aldehydes and ketones with the presence of amino acid transaminases [3, 4].

With specific relation to cheeses, the main detected BA are histamine, tyramine, putrescine, cadaverine and 2-phenylethylamine [5, 6]. These BA are products of the decarboxylation of histidine, tyrosine, ornithine, lysine and phenylalanine, respectively. Putrescine can also be obtained through deamination of agmatine [7].

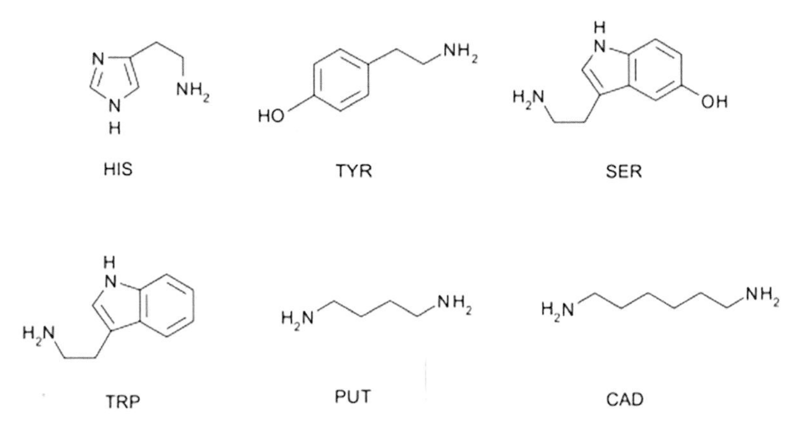

Fig. 1.1 Chemical structures of six biogenic amines: histamine (HIS), tyramine (TYR), serotonin (SER), tryptamine (TRP), putrescine (PUT), and cadaverine (CAD). BKchem version 0.13.0, 2009 (http://bkchem.zirael.org/index.html) has been used for drawing this structure

Decarboxylase-producing life forms can be found among the so-called starter culture, also named lactic acid bacteria (LAB), non-starter lactic acid bacteria (NS-LAB) and/or other spontaneous microflora [8, 9]. However, the capacity to produce biogenic amines is a strain-specific feature within microbial groups [10].

With relation to the aminogenesis process, the formation of BA depends on the activity of two proteins at least: a key enzyme with amino acid decarboxylase (AAD) activity, belonging to the pyridoxal-phosphate-dependent enzyme group (whose members use pyridoxal-5'-phosphate as a coenzyme) and a transporter molecule responsible for AA/BA interchange. These proteins belong to the amino acid/polyamine/organocation (APC) superfamily [11, 12].

BA production in foods requires the availability of precursors (i.e. AA), the presence of bacteria able to synthesise AAD and favourable conditions for their growth and decarboxylating activity [12, 13]. The formation of biogenic amines has been associated with some groups of microorganisms. Also, some strains are able to produce more than one amine at the same time, either due to the presence of different decarboxylases or to the action of a single enzyme which decarboxylates different AA [14].

AAD are enzymes produced in many microorganisms: these molecules can be naturally present in microbiota of food products or may be introduced by contamination before, during or after food processing. Decarboxylases have been found in species belonging to *Bacillus*, *Clostridium*, *Pseudomonas* and *Photobacterium*; they are also present in genera of the family *Enterobacteriaceae* (such as *Citrobacter*, *Klebsiella*, *Escherichia*, *Proteus*, *Salmonella* and *Shigella*) and *Micrococcaceae* (such as *Staphylococcus*, *Micrococcus* and *Kocuria*). On the other hand, many LAB like genera *Lactobacillus*, *Enterococcus*, *Carnobacterium*, *Pediococcus*, *Lactococcus* and *Leuconostoc* are capable to decarboxylate one or more AA [15, 16].

1.2 Analytical Determination of Biogenic Amines in Food Products

The interest in the presence and amount of BA in foodstuffs and beverages is always remarkable because these compounds play an important role in human health and food safety. In fact, these are involved in natural biological processes; nonetheless, BA may be hazardous for the human being if their levels in foods or beverages reach a critical threshold [17].

BA formation can occur during food processing and storage because of bacterial activities. Thus, higher amounts of certain amines may be found in foods as consequence of the use of poor quality raw materials, microbial contamination and inappropriate conditions during food processing and storage [1]. Consequently, biogenic amines have been suggested as chemical indicators of the hygienic quality of raw material and/or manufacturing practices. The quantity of BA can be considered a marker of the level of microbiological contamination in foods [7]. For these reasons and their potential toxicity, BA levels have to be monitored in food products [1, 2, 4, 13]. Since the presence of BA has great impact on food quality and safety, different methods have been developed for their identification and quantitative determination recently [15, 18].

The highest drawbacks in the analysis of BA in foods are, very often, their wide range of concentrations and the complexity of sampled matrices due to high protein and fat contents. These factors may complicate BA extraction [18, 19]. Thus, a large amount of compounds can interfere in the analyte signal, providing matrix effect to take into account. Matrix effects can be estimated on condition that slopes of calibration curves obtained from spiked samples before the extraction are considered, with standard solution at different concentrations (external calibration curves). A tolerable signal suppression or enhancement effect should be evaluated if the slope ratio (matrix/solvent) ranges between 0.8 and 1.2, whereas lower values than 0.8 or higher than 1.2 imply a strong matrix effect.

The quantification of biogenic amines is generally performed by means of chromatographic methods: thin layer chromatography, ion exchange chromatography, gas chromatography and high-performance liquid chromatography (HPLC). Biochemical assays and capillary electrophoresis have also been reported [18, 20, 21]. Currently, reversed-phase HPLC in combination with pre- or post-column derivatisation is the most widely used method [1]. Indeed, the European Food Safety Authority (EFSA) recommends HPLC as the best analytical system, in terms of reliability and sensitivity, for BA detection in fermented foods [1]. In general, sample clean-up and pre-concentration procedures, based on liquid–liquid or solid–phase extraction, can be applied to improve selectivity and sensitivity. Besides, pre- or post-column derivatisation is usually necessary to attain the desired sensitivity since the majority of BA display poor spectroscopic features. Typical labelling reagents used in AA analysis, such as *o*-phthaldialdehyde, fluorenylmethylchloroformate and dansyl- or dabsyl- chlorides have been adapted [18]. The determination involves usually acid extraction from a solid matrix, after the saturation and alkalinisation of the extract, followed by a liquid–liquid partition with an organic solvent. Solid samples are extracted with acidic solvents such as trichloroacetic, perchloric or hydrochloric acids [22]. The extraction of amines represents the critical step of the process and it affects negatively analytical recoveries. Furthermore, the selective extraction of single amines from the acid solution is a process influenced by many factors such as the type of acid, the type of organic solvent, the salt used for saturation, pH for amine extraction (liquid–liquid partition with the organic solvent), time and type of stirring, etc. In particular, the choice of acid has to be related to the characteristics of the peculiar food matrix. pH is the key parameter that certainly exerts the greatest influence on partition equilibrium of amines between the two phases (aqueous and organic). Since AA are the precursors for BA formation, analytical methods, aiming to determine both compounds, are thus of particular interest for two main reasons: reliable information about nutritional or hygienic quality, and immediate quality control monitoring during manufacturing steps. A particular example concerns cheese ripening (and the possibly associated BA formation) [18].

1.3 Physiological and Toxicological Aspects of the Main Biological Amines Found in Food Products

BA are involved in different physiological roles such as synaptic transmission, blood pressure control, hormonal mediation, allergic response and cellular growth control. Aromatic amines (tyramine, tryptamine and 2-phenylethylamine) show a vasoconstrictor action. Psychoactive amines such as dopamine and serotonin are neurotransmitters in the central nervous system while others (histamine and serotonin) present a vasodilator effect. Moreover, tyramine and histamine also act as hormonal mediators in humans and animals [2].

High amounts of certain amines constitute a potential public health concern due to their physiological and toxicological effects. The consumption of foods containing high BA amounts may cause several problems for susceptible consumers. Toxic effects ascribed to BA ingestion because of food contamination have been observed, particularly in individuals having dysfunctional biogenic amines-degrading mechanisms, either naturally or due to intake of alcohol or certain medications [1, 20]. Clinical signs appear normally between 30 min to a few hours after BA ingestion and disappear generally within 24 h. These symptoms are reported to be headaches, vertigo, nausea and vomiting, enhancing of secretion by gastric mucosa, gastrointestinal cramps, stomachache and diarrhoea, hypotension, tachycardia, extrasystoles, itching, nose congestion, rhinorrhoea, blepharitis, flush, pruritus and urticaria [20, 23]. The related intensity is dependent on the quantitative and qualitative kind of ingested amines [17]. For these reasons, the Food and Agriculture Organization of the United Nations (FAO) considers BA in foods as a biological hazard [24].

The EFSA considers tyramine and histamine as the most toxic BA but indicates that there is no available information to conduct quantitative BA risk assessment [1].

Among intoxications related with BA, the 'scombroid poisoning' caused by histamine ingestion is reported with a certain frequency. Similar incidents have been associated with the ingestion of spoiled fish of the Scombridae family (which gave rise to the term 'scombroid syndrome') [25, 26]. Cheese follows is notably associated with tyramine poisoning, the so-called 'Cheese reaction', due to the high content in aged cheese [27].

No legal limits for tyramine in foods have ever been set anywhere; in addition, there are no legal restrictions for histamine in other foods except for fish. It is generally assumed that histamine is the most toxic BA. However, later results [28] indicate that tyramine is even more toxic than histamine. Linares and co-workers [28] reported cytotoxicity threshold detected for histamine (441 mg kg^{-1}) and for tyramine (302 mg kg^{-1}). Nevertheless, concentrations found to be toxic in both cases are commonly found in BA-rich foods such as cheese [1, 29]. Concentrations below this threshold might also cause adverse reactions, although not so severe. However, people with less powerful BA-detoxification systems caused by genetic deficiencies, gastrointestinal disease, mono- or diamine oxidase inhibitor medication, or after ingestion of alcohol or other potentiating factors [30] might be at greater risk.

1.3.1 Histamine

HIS is synthesised by the pyridoxal phosphate containing *L*-histidine decarboxylase from AA histidine. It can be metabolised by oxidative deamination by means of diamine oxidase (DAO) [30].

The ability to produce HIS has been found in both Gram-negative and Gram-positive bacteria. Many Gram-negative bacteria normally found in contaminated foods are able to produce HIS. For example, strains of *Oenococcus oeni*, *Pediococcus parvalus* and *P. damnosus*, *Tetragenococcus* species, *Leuconostoc* species, *Lactobacillus hilgardii*, *L. buchnerii* and *L. curvatus* are known to produce HIS in fermented foods [31]. In recent years, HIS production has been occasionally described for some previously unrecognised species with reference to BA synthesis, such as *Streptococcus salivarius* subsp. *thermophilus* [32] and *Lactobacillus sakei* [33]. However, the proportion of decarboxylase-positive strains is low and their histaminogenic potential is weak or not proven in food.

HIS has important roles in human metabolism: it acts both as local hormone and as neurotransmitter. Moreover, mast cells, blood cells and neurons in the brain may contain HIS [34].

HIS modulates a variety of functions by means of the interaction with specific receptors on target cells, namely H_1, H_2 and H_3 receptors of the G-protein coupled receptor family [35]. Its physiological role includes gastric acid secretion, cell growth and differentiation, circadian rhythm, attention and cognition. HIS poisoning is the most common food-borne intoxication caused by BA. Although cheese has been implicated in outbreaks of HIS poisoning, such incidents have long been associated with the ingestion of spoiled fish [25, 26]. The intake of foods with high HIS concentration is associated with a range of toxicological effects. Typically, such poisoning appears with allergic reactions, acute gastrointestinal distress, headache, nasal secretion, bronchospasm, tachycardia, extrasystoles, hypotension, edema (eyelids), asthma and perhaps neurological and cutaneous symptoms. Intoxication is characterised by an incubation period ranging from a few minutes to hours, with symptoms that are usually noticeable for a few hours only. These symptoms can be particularly severe in individuals who are deficient in diamine oxidase—the HIS-degrading enzyme present in epithelial cells of the intestinal tract—for genetic or pharmacological reasons [30]. Health risks can increase if enzymatic systems are blocked by mono- or diamine oxidase inhibitors, gastrointestinal diseases, genetic deficiencies or potentiating factors such as alcohol and other biogenic amines [30].

A number of drugs containing DAO-inhibitors and alcohol are known to reduce DAO activity [36]. Individual differences in enzyme activities besides varying HIS concentrations in food may account for different tolerance levels. In healthy persons, dietary HIS is rapidly detoxified by amine oxidases, but they may develop severe intoxication symptoms because of high HIS amounts ingested with food such as scombroid fish or matured cheese [30, 37].

1.3.2 Tyramine and 2-Phenylethylamine

TYR and PHE are synthesised in humans from their corresponding AA (tyrosine and phenylalanine, respectively) by decarboxylation. Their catabolism is mainly mediated by monoamine oxidase (MAO).

The main TYR producer organisms found in fermented foods are Gram-positive bacteria within the genera *Leuconostoc*, *Lactococcus* [38, 39], *Enterococcus* such as *Enterococcus faecalis* and *E. faecium*, and *Lactobacillus* such as *Lactobacillus curvatus* and *L. brevis* [40, 41]. Also, PHE production is associated with TYR production as demonstrated for *Enterococcus spp.*, *L. curvatus* and *Staphylococcus spp.* [42].

Both amines have important physiological roles in heterotrophic cells. Physiological effects of TYR include peripheral vasoconstriction, increased cardiac output and respiration, elevated blood glucose and release of norepinephrine [41]. PHE is a physiological constituent in the mammalian brain: it has been linked with the regulation of mood and attention [44]; also, this amine can act as enhancer substance making easier the release of neurotransmitters catecholamine and serotonin [45]. The ingestion of high concentration of these BA might also causes toxic effects for human health, especially in individuals with reduced-MAO detoxifying activity. It has been hypothesised that dietary TYR and trace amines cause vasodilatation of the mesenteric vascular bed, increasing blood flow at the gastrointestinal level and thus facilitating their absorption [46]. Direct effects associated with specific receptors have also been reported at the cardiovascular level, causing an increase in heart rate. PHE ingestion has sometimes been associated with symptoms such as headache, dizziness and discomfort [47].

In addition, TYR and PHE have been proposed as the initiators of hypertensive crisis in certain patients and of dietary-induced migraine. Clinical signs appear between 30 min to a few hours following BA consumption and usually disappear within few hours; recovery is usually complete within 24 h [1]. TYR has also been considered a causative agent of certain food-induced migraines, together with PHE. However, migraine is a multifactorial problem not only affected by one component of the diet but also by other environmental, physiological and psychological factors.

1.3.3 Putrescine and Cadaverine

PUT and CAD are diamines that can be formed either as natural polyamines during de novo polyamine biosynthesis or as BA by decarboxylation [2]. PUT formation in animals and microorganisms requires the free amino acid ornithine and the enzyme ornithine decarboxylase. Alternatively, PUT can be produced from arginine via agmatine and carbamylputrescine [48]. AA lysine is decarboxylated by lysine decarboxylase to form CAD.

Different bacteria species exhibit decarboxylases with high activity. PUT and CAD production has mainly been related to Gram-negative bacteria, especially in the families *Enterobacteriaceae, Pseudomonadaceae and Shewanellaceae*, generally associated with spoilage [49]. *Enterobacteria* genera *Citrobacter, Klebsiella, Escherichia, Proteus, Salmonella* and *Shigella* are associated with the production of considerable PUT and CAD amounts in foods [50]. LAB, *lactobacilli* mainly, and *staphylococci* have been reported to be able to produce PUT and/or CAD [51].

PUT acts as a precursor for the physiological polyamines (i.e. spermidine and spermine) in humans, and all these molecules are involved in the regulation of cell growth, cell division and differentiation via the regulation of gene expression and the modulation of signal transduction pathways [8, 35].

PUT is an important constituent of all mammalian cells and is essentially involved in a variety of regulatory steps during normal and malignant cell proliferation. PUT and CAD may react with nitrite to form carcinogenic nitrosamines; they can also be proposed as spoilage indicators [52]. Probably, the most relevant issue of PUT and CAD is related with their role as enhancers of other BA toxic effects due to the inhibition of detoxifying enzymes [53, 54]. In fact, PUT and CAD may potentiate toxic effects of HIS and TYR by inhibiting monoamine oxidase, diamine oxidase and hydroxymethyl transferase [2].

1.4 General Legislation Concerning Biogenic Amines in Foods

Legislation on BA in foods appears insufficient although it is generally accepted that BA should not be allowed to accumulate in food products. It is very difficult to establish a uniform maximum limit for ingested BA since their toxic effect depends on the BA typology, the presence of modulating compounds and the efficiency of each person's detoxification system.

The only BA for which maximum limits have been legally set by the EFSA is HIS, and then only in scombroid-like fish (200 mg kg^{-1}) and fish products (400 mg kg^{-1}) [55]. The United States Food and Drug Administration (FDA) has suggested that HIS concentration in fish >500 mg kg^{-1} correspond to health dangers [25, 56]. No further legislation exists anywhere to regulate HIS or the remaining BA in any other food.

1.5 Presence of Biogenic Amines in Cheese

After fish, cheese is the next most commonly implicated food item associated with BA poisoning [12]. Fermented dairy products and cheeses, in particular, may be a significant BA source [7, 12, 57]. High concentrations of HIS, TYR, PUT, CAD and PHE [4, 9] have been found in fermented foods such as cheeses; the phenomenon is ascribed to responsible enzymes—AAD—and microorganisms, for example in naturally occurring and/or artificially added LAB, involved in food fermentation [58].

Some strains described as part of the main amine producers in cheeses are present in the genera *Enterococcus, Lactobacillus, Lactococcus, Leuconostoc, Streptococcus* and the *Enterobacteriaceae* family [12].

Cheese is an ideal substrate for amines production. Indeed, the main biochemical process (proteolysis) that takes place during cheese ripening leads to the accumulation of free AA, and some of these are BA precursors. As a consequence, the higher the content of free AA, the higher the probability of BA production [4, 57, 59]. Other factors affecting BA production in cheeses include technological processes with the possible presence of decarboxylase-positive microorganisms and the synergistic effects between microorganisms as well [4, 60], but also environmental conditions that allow their growth, as well as the presence of suitable cofactors (pyridoxal phosphate) [12]. For these reasons and their potential toxicity, several authors studied BA concentration values in different kinds of cheese [1, 4–6, 9, 19, 22, 61, 62]. Obtained data allow assuming that the highest BA amounts are present in hard-ripened cheese. These compounds can be present in all samples with notably wide concentration ranges because BA production and the accumulation in fermented foods is an extremely complex phenomenon affected by multiple factors and their interactions [12, 22].

1.6 Biogenic Amines in Cheeses: Technological Aspects

BA concentration and types in cheeses are extremely variable. In particular [1, 4–6, 9, 19, 22, 61, 62]:

(a) PUT can be detected between 4 and 298 mg kg^{-1} depending on the typology of cheese and ripening periods. Interestingly, the maximum amount is found in ripened cheeses while hard-ripened cheeses are reported to contain slightly lower quantities (up to 204 mg kg^{-1})
(b) HIS may reach 646 mg kg^{-1} in hard-ripened products
(c) TYR can be considered the higher concern with relation to maximum amounts in hard-ripened cheeses, although the problem can be signalled in ripened cheeses also (maximum reported quantity: 1009 mg kg^{-1} in hard-ripened cheeses)
(d) PHE seems to be found in reduced quantities when speaking of ripened, hard-ripened and unripened cheeses (the maximum detection does not appear to exceed 145 mg kg^{-1}).

Aminogenesis processes depend on multiple and complex variables such as type and thermal treatment of milk, presence of microorganisms and related proteolytic and decarboxylase activities, ripening time, pH and sodium chloride (NaCl) concentration, storage time and temperature and packaging type [22]. In addition, the availability of substrate AA is one of the prerequisites for BA synthesis. In particular, proteolysis is a crucial factor because it is directly related to availability of free AA. It has been reported that conditions of accelerated or enhanced proteolysis increase BA formation [38, 57].

1.6.1 Ripening Time

One of the main factors strongly affecting BA accumulation in cheese is the ripening period. Cheese ripening is a complex phenomenon, producing a series of physical-chemical, biochemical and biologic variations strictly connected. In this process, caseins represent the substrate but lipids and soluble components (sugar, lactic acid, citric acid, etc.) are involved as well. Cheese proteolysis increases AA availability; AA can be decarboxylated by microbial enzymes to produce BA. Moreover, the use of starter cultures could influence BA production, either directly or indirectly, through the interaction with the wild flora [13, 63].

During fermentation, and besides the contribution of contaminating bacteria, microbiota responsible for fermentation can also show aminogenic activity. Moreover, proteolytic activity and other biochemical mechanisms such as yeast lysis and acidification, usually accompanying fermentation processes, increase the availability of precursor free AA with favoured decarboxylation reactions [13]. However, microorganisms must be carefully selected for each type of product and variety (i.e. substrate) taking into account their technological competence (competitiveness, influence on organoleptic characteristics, etc.) and safety requirements, including the inability to produce BA, which is not a usual criterion for starter culture selection [64].

Caseins represent the substrate but the process involves also lipids and soluble components (sugar, lactic acid, citric acid, etc.). Starter LAB and NS-LAB and/or other spontaneous microflora carry out the ripening process. Starter LAB contribute to protein degradation, while NS-LAB are responsible for peptidolysis and BA release [65]. Free AA can be further decarboxylated by bacterial enzymes, producing BA accumulation. Therefore, prolonged ripening times are considered one of the main causes of BA augments in cheese, as reported by several authors [6, 63, 64]. It can be affirmed that BA contents in ripened cheeses are generally higher than those in unripened cheeses.

Moreover, great variations have been reported in the type and quantity of produced BA between different strains of the same species or genera [9, 48]. In many cases, BA accumulation has been attributed mainly to the activity of the non-starter microflora. However, an indirect role of the starter LAB can be hypothesised: peptidases released by starter LAB lysis could be essential to provide precursor AA [54]. It follows that, during the fermentation process, BA formation can be controlled by using starter cultures that are less effective in decarboxylating AA; the inability to produce BA should be considered a necessary condition for strains intended to be used as starters. Another approach to reduce BA in cheese is to use starter cultures with amine oxidase able to degrade BA [66].

1.6.2 Kind of Milk

BA content in cheeses depends mainly on the microbiological quality of milk if compared with the milk typology. However, many authors found that thermal treatment is able to reduce BA concentration in cheeses. In fact, many decarboxylating bacteria do not survive at high temperature [7, 64, 67]. Other factors should be considered, including the slower rate of proteolysis observed in cheeses made from pasteurised milk and heat inactivation of pyridoxal phosphate, the cofactor for decarboxylase activity [68]. This is the main explanation for lower BA contents normally detected in cheeses from pasteurised milk or from milk subjected to high-pressure homogenisation in comparison with those obtained from raw milk [63, 68]. Anyway, deficient hygienic conditions can promote contamination with BA-producing microorganisms during cheese manufacturing after the thermal treatment, leading to BA accumulation (remarkable amounts) in cheeses derived from pasteurised milk as well [67, 61].

1.6.3 pH and NaCl Concentration

pH level is an important parameter because of the related influence on AAD activity [4]. Generally, the pH of cheeses varies from 4.0 to 6.0 [22] and AA decarboxylating enzymes show optimum activity at acid pH values (4.0–5.5) [38]. In particular, BA formation could be influenced from two main pH-dependent mechanisms acting simultaneously.

First, the acidification could inhibit the growth of decarboxylation microorganisms [41]. In addition, the production and enzymatic activity may be influenced at low pH values because bacteria are more stimulated to produce decarboxylase as a part of their defence mechanisms against acidity [38, 40]. Salt concentration can influence cheese texture, flavour development and BA formation from free AA. Different salt contents could be related with the variation of microflora composition, leading to differences in BA formation. Some studies found that the high salt content could control BA production [66]. This observation could be explained by means of the inhibitory effect of high salt content on the growth rate of BA-producing bacteria [70].

1.6.4 Storage Time and Temperature

Storage temperature is an important technological parameter strongly affecting BA formation in cheeses. Several authors have found that the quantitative BA production is usually reported to be temperature- and time-dependent [5, 7, 66]. Mostly, amine production rates increase with the temperature. In fact, microbial growth, biochemical reactions and the production of metabolites increase generally with temperature augments. Conversely, BA accumulation is minimised at low temperatures, through inhibition of microbial growth and the reduction of enzymatic activity [22, 66]. The optimum temperature for BA production by mesophilic bacteria has been reported to be between 20 and 37 °C, while related decreases are observed below 5 °C or above 40 °C [1]. For these reasons, low temperatures should be applied during storage to reduce proteolytic, decarboxylase activities and microbial growth [71].

1.6.5 Packaging

Modified atmosphere packaging (MAP) is a popular food preservation method involving the modification of gas composition, surrounding the food product and packaging with barrier films [72]. Combination of carbon dioxide (CO_2), nitrogen and oxygen (O_2) in MAP-perishable foods is able to suppress the aerobic spoilage flora [73]. Some authors [50, 73, 74] reported studies on the successful control of

BA formation through MAP in some foods [66]. All these reports represent BA reduction with MAP techniques but the effectiveness of the method depends on gas concentration, microbial ecology of the food and storage conditions. Andic and co-workers [74] have investigated the effect of packaging (non-vacuum and vacuum) methods on BA production in cheese during storage. It has been found that packaging method and storage temperature (4 and 18 °C) had a significant effect on BA concentrations. In particular, they reported that non-vacuum-packaged cheeses stored at 4 °C had higher PUT, CAD, HIS and TYR amounts in comparison with vacuum-packaged cheeses.

1.6.6 Possible Mitigation Strategies Against Biogenic Amines

Safety is a basic requirement in food productions. At present, legislation defining BA limits and related tolerances in fermented foodstuff does not appear sufficient. However, cheese should be strictly monitored because of related high BA concentrations. Therefore, a greater knowledge of the factors involved in BA synthesis and accumulation should lead to a reduction in their incidence in dairy foods.

As above mentioned, the main prerequisites for BA production in foods include: availability of free AA, presence of BA-producing microorganisms, thermal treatment of milk, ripening time, pH and NaCl concentration, type of packaging, storage time and temperature. However, many physical-chemical parameters cannot be easily modified as they are closely related to the fermentation process.

AA availability, an important factor related to BA production, may be difficultly reduced in dairy products since proteolysis is essential in cheese ripening, and AA are required for flavour development [75]. This reflection supports the idea that the control of BA-producing microorganisms by means of adequate milk treatments is one of the most important factors for BA minimisation in dairy products [76]. This result could be achieved by means of the application of thermal treatments to facilitate the dominance of starter bacteria from the early stages of fermentation and by means of the selection of non-BA producers (these cultures should be able to outgrow autochthonous microbiota under production conditions in the industry [58, 66]).

The selection of starters without BA synthesis capability is an important mitigation strategy for BA reduction in dairy products. Anyway, it should be noted that the growing interest in artisanal cheeses is partly due to the uniqueness of such products, in which specialised microorganisms can grow and contribute to their organoleptic and qualitative features [59]. However, these dairy products are often manufactured under poor or uncontrolled hygiene conditions; in addition, they are produced following different protocols, which can vary from one to another cheesemaker. Many cheesemakers use raw milk for sensorial reasons (enhanced or stronger flavour) instead of pasteurised milk, primarily due to greater proteolysis and lipolysis by raw milk microbiota in the cheese [77]. These microorganisms play a major role in the development of the organoleptic characteristics of cheeses.

References

1. EFSA (2011) Scientific opinion on risk based control of biogenic amine formation in fermented foods. Panel on Biological Hazards (BIOHAZ). EFSA J 9, 10:2393–2486. doi:10.2903/j.efsa.2011.2393
2. Bardocz S (1995) Polyamines in food and their consequences for food quality and human health. Trends Food Sci Technol 6(10):341–346. doi:10.1016/S0924-2244(00)89169-4
3. Zolou A, Lokou Z, Soufleros E, Stratis I (2003) Determination of biogenic amines in wines and beers by high performance liquid chromatography with pre-column dansylation and ultraviolet detection. Chromatographia 57(7–8):429–439. doi:10.1007/BF02492537
4. Spizzirri UG, Restuccia D, Curcio M, Parisi OI, Iemma F, Picci N (2013) Determination of biogenic amines in different cheese samples by LC with evaporative light scattering detector. J Food Compost Anal 29(1):43–51. doi:10.1016/j.jfca.2012.09.005
5. Linares DM, Del Rio B, Ladero V, Martínez N, Fernández M, Martín MC, Álvarez MA (2012) Factors influencing biogenic amines accumulation in dairy products. Front Microbiol 3 (180):1–10. doi:10.3389/fmicb.2012.00180
6. Buňková L, Buňka F, Klčovská P, Mrkvička V, Doležalová M, Kráčmar S (2010) Formation of biogenic amines by Gram-negative bacteria isolated from poultry skin. Food Chem 121 (1):203–206. doi:10.1016/j.foodchem.2009.12.012
7. Ercan SS, Bozkurt H, Soysal Ç (2013) Significance of biogenic amines in foods and their reduction methods. J Food Sci Eng 3(8):395–410
8. Halász A, Baráth Á, Simon-Sarkadi L, Holzapfel W (1994) Biogenic amines and their production by micro-organisms in food. Trends Food Sci Technol 5, 2:42–49. doi:10.1016/0924-2244(94)90070-1
9. Guarcello R, Diviccaro A, Barbera M, Giancippoli E, Settanni L, Minervini F, Moschetti G, Gobbetti M (2015) A survey of the main technology, biochemical and microbiological features influencing the concentration of biogenic amines of twenty Apulian and Sicilian (Southern Italy) cheeses. Int Dairy J 43:61–69. doi:10.1016/j.idairyj.2014.11.007
10. Coton E, Coton M (2009) Evidence of horizontal transfer as origin of strain to strain variation of the tyramine production trait in *Lactobacillus brevis*. Food Microbiol 26(1):52–57. doi:10.1016/j.fm.2008.07.009
11. Jack DL, Paulsen IT, Saier MH (2000) The amino acid/polyamine/organocation (APC) superfamily of transporters specific for amino acids, polyamines and organocations. Microbiol 146(8):1797–1814. doi:10.1099/00221287-146-8-1797
12. Linares DM, Martín M, Ladero V, Alvarez MA, Fernández M (2011) Biogenic amines in dairy products. Crit Rev Food Sci Nutr 51(7):691–703. doi:10.1080/10408398.2011.582813
13. Ten Brink B, Damink C, Joosten HMLJ, Huis in't Veld JHJ (1990) Occurrence and formation of biologically active amines in foods. Int J Food Microbiol 11(1):73–84. doi:10.1016/0168-1605(90)90040-C
14. Bover-Cid S, Holzapfel W (1999) Improved screening procedure for biogenic amine production by lactic acid bacteria. Int J Food Microbiol 53(1):33–41. doi:10.1016/S0168-1605(99)00152-X
15. Karovičová J, Kohajdová Z (2005) Biogenic amines in food. Chem Pap 59(1):70–79
16. Galgano F, Favati F, Bonadio M, Lorusso V, Romano P (2009) Role of biogenic amines as index of freshness in beef meat packed with different biopolymeric materials. Food Res Int 42 (8):1147–1152. doi:10.1016/j.foodres.2009.05.012

17. Ladero V, Calles-Enríquez M, Fernández M, Alvarez AM (2010) Toxicological effects of dietary biogenic amines. Curr Nutr Food Sci 6(2):145–156. doi:10.2174/157340110791233256
18. Önal A, Tekkeli S, Önal C (2013) A review of the liquid chromatographic methods for the determination of biogenic amines in foods. Food Chem 138(1):509–515. doi:10.1016/j.foodchem.2012.10.056
19. Gosetti F, Mazzucco E, Gianotti V, Polati S, Gennaro MC (2007) High performance liquid chromatography/tandem mass spectrometry determination of biogenic amines in typical Piedmont cheeses. J Chromatogr A 1149(2):151–157. doi:10.1016/j.chroma.2007.02.097
20. Önal A (2007) A review: current analytical methods for the determination of biogenic amines in foods. Food Chem 103(4):1475–1486. doi:10.1016/j.foodchem.2006.08.028
21. Rodriguez MBR, Da Silva Carneiro C, Da Silva Feijó MB, Rodriguez MBR, da Silva Carneiro C, da Silva Feijó MB, Júnior CAC, Mano SB (2014) Bioactive amines: aspects of quality and safety in food. Food Nutr Sci 5(2):138–146. doi:10.4236/fns.2014.52018
22. Loizzo MR, Menichini F, Picci N, Puoci F, Spizzirri UG, Restuccia D (2013) Technological aspects and analytical determination of biogenic amines in cheese. Trends Food Sci Technol 30(1):38–55. doi:10.1016/j.tifs.2012.11.005
23. Yoon H, Park JH, Choi A, Hwang HJ, Mah JH (2015) Validation of an HPLC analytical method for determination of biogenic amines in agricultural products and monitoring of biogenic amines in korean fermented agricultural products. Toxicol Res 31(3):299–305. doi:10.5487/TR.2015.31.3.299
24. FAO (2014) Assessment and management of seafood safety and quality. Current practices and emerging issues. FAO Fisheries and Aquaculture Technical Paper 574. Food and Agriculture Organization of the United Nations, Rome
25. FDA (2011) Fish and Fishery Products - Hazards and Controls Guidance, Fourth Edition. Washington, DC: U.S. Department of Health and Human Services, Food and Drug Administration (FDA), Center for Food Safety and Applied Nutrition, Office of Seafood, Washington, D.C.
26. Visciano P, Schirone M, Tofalo R, Suzzi G (2014) Histamine poisoning and control measures in fish and fishery products. Front Microbiol 5(500):1–3. doi:10.3389/fmicb.2014.00500
27. Schirone M, Tofalo R, Visciano P, Corsetti A, Suzzi G (2012) Biogenic amines in Italian Pecorino cheese. Front Microbiol 3:171. doi:10.3389/fmicb.2012.00171
28. Linares DM, Del Rio B, Redruello B, Ladero V, Cruz Martin M, Fernandez M, Ruas-Madiedo P, Alvarez MA (2016) Comparative analysis of the in vitro cytotoxicity of the dietary biogenic amines tyramine and histamine. Food Chem 197 A:658–663. doi:10.1016/j.foodchem.2015.11.013
29. Redruello B, Ladero V, Cuesta I, Álvarez-Buylla JR, Martín MC, Fernández M, Alvarez MA (2013) A fast, reliable, ultra high performance liquid chromatography method for the simultaneous determination of amino acids, biogenic amines and ammonium ions in cheese, using diethyl ethoxymethylenemalonate as a derivatising agent. Food Chem 139(1–4):1029–1035. doi:10.1016/j.foodchem.2013.01.071
30. Maintz L, Novak N (2007) Histamine and histamine intolerance. Am J Clin Nutrit 85 (5):1185–1196
31. Spano G, Russo P, Lonvaud-Funel A, Lucas P, Alexandre H, Grandvalet C, Coton E, Coton M, Barnavon L, Bach B, Rattray F, Bunte A, Magni C, Ladero V, Alvarez M, Fernández M, Lopez P, de Palencia PF, Corbi A, Trip H, Lolkema JS (2010) Risk assessment of biogenic amines in fermented food. Eur J Clin Nutr 64:S95–S100. doi:10.1038/ejcn.2010.218
32. Calles-Enríquez M, Eriksen BH, Andersen PS, Calles-Enríquez M, Eriksen BH, Andersen PS, Rattray FP, Johansen AH, Fernández M, Ladero V, Alvarez MA (2010) Sequencing and Transcriptional Analysis of the *Streptococcus thermophilus* histamine biosynthesis gene cluster: Factors that affect differential hdcA Expression. Appl Environ Microbiol 76 (18):6231–6238. doi:10.1128/AEM.00827-10

33. Coton E, Coton M (2005) Multiplex PCR for colony direct detection of Gram-positive histamineand tyramine-producing bacteria. J Microbiol Meth 63(3):296–304. doi:10.1016/j.mimet.2005.04.001
34. Forth W, Henschler D, Rummel W, Starke K (2001) Allgemeine und spezielle Pharmakologie und Toxikologie, 8th edn. Urban & Fischer Verlag, München and Jena
35. Shukla S, Kim JK, Kim M (2011) Occurrence of biogenic amines in soybean food products. In: El-Shemy H (ed) Soybean and Health. InTech, Rijeka and Shangai
36. Sattler J, Häfner D, Klotter HJ, Lorenz W, Wagner PK (1988) Food induced histaminosis as an epidemiological problem: plasma histamine elevation and haemodynamic alterations after oral histamine administration and blockade of diamine oxidase (DAO). Agents Actions 23(3–4):361–365. doi:10.1007/BF02142588
37. Jarisch R, Götz M, Hemmer W, Missbichler A, Raithel M, Wantke F (2004) Histamin-Intoleranz. Histamin und Seekrankheit, vol 2. Georg Thieme Verlag, Stuttgart and New York
38. Fernández M, Linares DM, Rodríguez A, Alvarez MA (2007) Factors affecting tyramine production in *Enterococcus durans* IPLA 655. Appl Microbiol Biotechnol 73(6):1400–1406. doi:10.1007/s00253-006-0596-y
39. Fernández M, Linares DM, del Río B, Ladero V, Alvarez MA (2007) HPLC quantification of biogenic amines in cheeses: correlation with PCR-detection of tyramine-producing microorganisms. J Dairy Res 74(3):276–282. doi:10.1017/S0022029907002488
40. Bover-Cid S, Hugas M, Izquierdo-Pulido M, Vidal-Carou MC (2000) Amino acid-decarboxylase activity of bacteria isolated from fermented pork sausages. Int J Food Microbiol 66(3):185–189. doi:10.1016/S0168-1605(00)00526-2
41. Bover-Cid S, Hugas M, Izquierdo-Pulido M, Vidal-Carou MC (2000) Reduction of biogenic amine formation using a negative amino acid-decarboxylase starter culture for fermentation of fuet sausage. J Food Protect 63(2):237–243. doi:10.4315/0362-028X-63.2.237
42. Suzzi G, Gardini F (2003) Biogenic amines in dry fermented sausages: a review. Int J Food Microbiol 88(1):41–54. doi:10.1016/S0168-1605(03)00080-1
43. Marcobal A, De Las Rivas B, Landete JM, Tabera L, Muñoz R (2012) Tyramine and phenylethylamine biosynthesis by food bacteria. Crit Rev Food Sci Nutr 52(5):448–467. doi:10.1080/10408398.2010.500545
44. Szabo A, Billet E, Turner J (2001) Phenylethylamine, a possible link to the antidepressant effects of exercise. Br J Sports Med 35(5):342–343. doi:10.1136/bjsm.35.5.342
45. Shimazu S, Miklya I (2004) Pharmacological studies with endogenous enhancer substances: Beta-phenylethylamine, tryptamine, and their synthetic derivatives. Prog Neuropsychopharmacol Biol Psychiatry 28(3):421–427. doi:10.1016/j.pnpbp.2003.11.016
46. Broadley KH (2010) The vascular effects of trace amines and amphetamines. J Pharmacol Exp Ther 125(3):363–375. doi:10.1016/j.pharmthera.2009.11.005
47. Frascarelli S, Ghelardoni S, Chiellini G, Vargiu R, Ronca-Testoni S, Scanlan TS, Grandy DK, Zucchi R (2008) Cardiac effects of trace amines: pharmacological characterization of trace amine-associated receptors. Eur J Pharmacol 587(1–3):231–236. doi:10.1016/j.ejphar.2008.03.055
48. Lucas PM, Blancato VS, Claisse O, Magni C, Lolkema JS, Lonvaud-Funel A (2007) Agmatine deiminase pathway genes in *Lactobacillus brevis* are linked to the tyrosine decarboxylation operon in a putative acid resistance locus. Microbiol 153:2221–2230. doi:10.1099/mic.0.2007/006320-0
49. López-Caballero ME, Sánchez-Fernández JA, Moral A (2001) Growth and metabolic activity of *Shewanella putrefaciens* maintained under different CO_2 and O_2 concentrations. Int J Food Microbiol 64(3):277–287. doi:10.1016/S0168-1605(00)00473-6
50. Kim MK, Mah JH, Hwang HJ (2009) Biogenic amine formation and bacterial contribution in fish, squid and shellfish. Food Chem 116(1):87–95. doi:10.1016/j.foodchem.2009.02.010
51. Arena ME, Manca de Nadra MC (2001) Biogenic amine production by *Lactobacillus*. J Appl Microbiol 90(2):158–162. doi:10.1046/j.1365-2672.2001.01223.x

52. Hernández-Jover T, Izquierdo-Pulido M, Veciana-Nogués MT, Mariné-Font A, Vidal-Carou MC (1997) Biogenic amine and polyamine contents in meat and meat products. J Agric Food Chem 45(6):2098–2102. doi:10.1021/jf960790p
53. Shalaby AR (1996) Significance of biogenic amines to food safety and human health. Food Res Int 29(7):675–690. doi:10.1016/S0963-9969(96)00066-X
54. Valsamaki K, Michaelidou A, Polychroniadou A (2000) Biogenic amine production in Feta cheese. Food Chem 71(2):259–266. doi:10.1016/S0308-8146(00)00168-0
55. Comission European (2005) Commission Regulation (EC) No. 2073/2005 of 15 November 2005 on microbiological criteria for foodstuffs. Off J Eur Union L338:1–26
56. FDA (1996) Decomposition and histamine in raw, frozen tuna and mahi-mahi canned tuna; and related species. Compliance Policy Guides 7108.240, Section 540.525. Food and Drug Administration (FDA), Rockville
57. Innocente N, D'Agostin P (2002) Formation of biogenic amines in a typical semihard Italian cheese. J Food Prot 65(9):1498–1501. doi:10.4315/0362-028X-65.9.1498
58. Kalac P (2009) Recent advances in the research on biological roles of dietary polyamines in man. J Appl Biomed 7:65–74
59. Schirone M, Tofalo R, Fasoli G, Perpetuini G, Corsetti A, Manetta AC, Ciarrocchi A, Suzzi G (2013) High content of biogenic amines in Pecorino cheeses. Food Microbiol 34(1):137–144. doi:10.1016/j.fm.2012.11.022
60. Marino M, Maifreni V, Moret S, Rondinini G (2000) The capacity of *Enterobacteriaceae* species biogenic amines in cheese. Lett Appl Microbiol 31(2):169–173. doi:10.1046/j.1365-2672.2000.00783.x
61. Ladero V, Fernández M, Alvarez MA (2009) Effect of post-ripening processing on the histamine and histamine-producing bacteria contents of different cheeses. Int Dairy J 19 (12):759–762. doi:10.1016/j.idairyj.2009.05.010
62. Mayer HK, Fiechter G, Fischer E (2010) A new ultra-pressure liquid chromatography method for the determination of biogenic amines in cheese. J Chromatogr A 1217, 19:3251–3257. doi:10.1016/j.chroma.2009.09.027
63. Ordóñez AI, Ibáñez FC, Torre P, Barcina Y (1997) Formation of biogenic amines in Idiazábal ewe's-milk cheese: effect of ripening, pasteurization, and starter. J Food Prot 60(11):1371–1375. doi:10.4315/0362-028X-60.11.1371
64. Holzapfel WH (2002) Appropriate starter culture technologies for small-scale fermentation in developing countries. Int J Food Microbiol 75(3):197–212. doi:10.1016/S0168-1605(01)00707-3
65. Martuscelli M, Gardini F, Torriani S, Mastrocola D, Serio A, Chaves-López C, Schirone M, Suzzi G (2005) Production of biogenic amines during the ripening of Pecorino Abruzzese cheese. Int Dairy J 15, 6:571–578. doi:10.1016/j.idairyj.2004.11.008
66. Chong CY, Abu Bakar F, Abdul Rahman R, Bakar J, Mahyudin NA (2011) Mini review the effects of food processing on biogenic amines formation. Int Food Res J 18(3):867–876
67. Lanciotti R, Patrignani F, Iucci L, Guerzoni ME, Suzzi G, Belletti N, Gardini F (2007) Effects of milk high pressure homogenization on biogenic amine accumulation during ripening of ovine and bovine Italian cheeses. Food Chem 104(2):693–701. doi:10.1016/j.foodchem.2006.12.017
68. Ladero V, Linares DM, Fernández M, Alvarez MA (2008) Real time quantitative PCR detection of histamine-producing lactic acid bacteria in cheese: relation with histamine content. Food Res Int 41(10):1015–1019. doi:10.1016/j.foodres.2008.08.001
69. Gardini F, Martuscelli M, Caruso MC, Galgano F, Crudele MA, Favati F, Guerzoni ME, Suzzi G (2001) Effects of pH, temperature and NaCl concentration on the growth kinetics, proteolytic activity and biogenic amine production of Enterococcus faecalis. Int J Food Microbiol 64(1):105–117. doi:10.1016/S0168-1605(00)00445-1
70. Rezaei M, Montazeri N, Langrudi HE, Mokhayer B, Parviz M, Nazarinia A (2007) The biogenic amines and bacterial changes of farmed rainbow trout (*Oncorhynchus mykiss*) stored in ice. Food Chem 103(1):150–154. doi:10.1016/j.foodchem.2006.05.066

71. Jeremiah LE (2001) Packaging alternatives to deliver fresh meats using short- or long-term distribution. Food Res Int 34(9):749–772. doi:10.1016/S0963-9969(01)00096-5
72. Emborg J, Laursen BG, Dalgaard P (2005) Significant histamine formation in tuna (Thunnus albacares) at 2 °C—Effect of vacuum and modified atmosphere packaging on psychrotolerant bacteria. Int J Food Microb 101(3):263–279. doi:10.1016/j.ijfoodmicro.2004.12.001
73. Patsias A, Chouliara I, Paleologos EK, Savvaidis I, Kontominas MG (2006) Relation of biogenic amines to microbial and sensory changes of precooked chicken meat stored aerobically and under modified atmosphere packaging at 4 °C. Eur Food Res Techn 223 (5):683–689. doi:10.1007/s00217-006-0253-9
74. Andiç S, Gençcelep H, Tunçtürk Y, Köse Ş (2010) The effect of storage temperatures and packaging methods on properties of Motal cheese. J Dairy Sci 93(3):849–859. doi:10.3168/jds.2009-2413
75. Fernández M, Zúñiga M (2006) Amino acid catabolic pathways of lactic acid bacteria. Crit Rev Microbiol 32(3):155–183. doi:10.1080/10408410600880643
76. Novella-Rodríguez S, Veciana-Nogués MT, Roig-Sagués AX, Trujillo-Mesa AJ, Vidal-Carou MC (2004) Evaluation of biogenic amines and microbial counts throughout the ripening of goat cheeses from pasteurized and raw milk. J Dairy Res 71(2):245–252. doi:10.1017/S0022029904000147
77. Buffa MN, Trujillo AJ, Pavia Guamis B (2001) Changes in textural, microstructural, and colour characteristics during ripening of cheeses made from raw, pasteurized or high-pressure-treated goats' milk. Int Dairy J 11(11):927–934. doi:10.1016/S0958-6946(01)00141-8

Chapter 2
Evolutive Profiles of Caseins and Degraded Proteins in Industrial High-Moisture Mozzarella Cheeses. A Simulative Approach

Abstract This Chapter evaluates the consequences of protein modifications in cheeses, with special emphasis on mozzarella cheeses. Basic features of modern and historically relevant cheeses depend on the chemical and physical state of main components. The palatability and consumeristic acceptance strongly depend on the flavour and taste features of the fat phase in foods. On the other side, the modification of proteins is interesting. With specific reference to caseins, the main nitrogen-based structure of the final cheese product, many factors influence protein degradation. Because of the needed bioavailability of free water, the amount of moisture becomes important enough. Unfortunately, higher moisture contents may mean lower shelf-life values and enhanced proteolytic degradation. The aim of this Chapter has been to show analytical results of an emulation study carried out on different industrial high-moisture mozzarella cheeses during storage. Obtained data and calculated results seem to suggest that the amount of small molecules increases globally during time, but demolition is mainly ascribed to medium and large protein molecules.

Keywords Absorption · Casein · Cow's milk · CYPEP:2006 · Refrigerated storage · Low-moisture mozzarella cheese · Moisture · Emulation

Abbreviations

A%	Apparent hydric absorption
CYPEP:2006	Cheesemaking Yield and Proteins Estimation according to Parisi:2006
DP	Degraded protein
FC	Fat matter
HAC	High-absorption casein
HMM	High-moisture mozzarella
MC	Moisture
MW	Molecular weight
PA	Protein
SimP	Simulated polypeptide

© The Author(s) 2018

C. Barone et al., *Chemical Evolution of Nitrogen-Compounds in Mozzarella Cheeses*, SpringerBriefs in Molecular Science,
DOI 10.1007/978-3-319-65739-4_2

2.1 Cheese Proteins and Caseins. Chemical Degradation

Cheese is an ideal substrate for amines production, because of the tripartite composition of this product and its microbial ecology [1–4]. In fact, each typology of cheese from goat, cow, or sheep milks is approximately represented with a three-components formulation concerning moisture content (it is the estimated water amount in the product), lipids (also named 'fat matter'), and proteins, with a minor presence of carbohydrates (residual lactose is the main fraction) and salt (sodium chloride, calcium salts) [5, 6]. Cheeses may be considered the result of the dissolution of solid compounds in aqueous media. On the other side, solid compounds—lipids, proteins, salts, carbohydrates, etc.,—correspond to the solid matter of cheeses, and are generally defined 'dry matter'. Fat matter and non-protein molecules are trapped into a solid matrix made of proteins with various molecular weights and different spatial arrangements.

The comprehension of cheese structures may be more complicated than the above-described system because of the nature of proteins in cheeses. These proteins, with variable dimensions and physical-chemical features depending on the peculiar type of milk (cow milk is the most used raw matter, but other milks can be considered), should be further subdivided in three categories:

(a) Caseins (with relation to cow milk, four different casein molecules are known and described)
(b) Lactoglobulins
(c) Lactoalbumins.

In general, the production of cheeses is based on the precipitation of a 'caseous' matrix from the original milk at acid pH values, trapping all other solid substances with certain molecular weights. This simple descriptions highlights the role of 'caseins', while other proteins are not mentioned explicitly [7]. In fact, caseins only can be modified chemically and physically in this way in synergy with water (because of the formation of different hydrogen bonds). On the other side, lactoalbumins and lactoglobulins cannot coagulate at acid pH values; the result of initial cheese productions is always the intermediate agglomeration of caseins and other solid substances (with the exclusion of low-sized molecules) with a considerable water amount. On the other hand, remaining proteins are 'filtered off' (actually, expelled) from the new caseous matrix with residual carbohydrates (too low molecular weights) and different organic or inorganic compounds.

Basic features of modern and historically relevant cheeses depend on the chemical and physical state of main components: Certainly, the modification of fat matter after lipolysis and other chemical mechanisms is important by the marketing and technological viewpoints at least. The palatability and consumeristic acceptance strongly depend on the flavour and taste features of the fat phase in foods [7]. On the other side, the modification of proteins may appear more interesting because of direct and indirect influences on the yield of cheesemaking procedures and the safety (hygienic profiles) of the final product and the related process.

With specific reference to caseins, the main nitrogen-based structure of the final cheese product, many factors influence protein degradation. Generally, these reactions are ascribed to microbial life forms: proteolytic microorganisms. Because of the needed bioavailability of free water, the amount of moisture becomes important enough. When speaking of cheesemaking yields, the higher the moisture, the higher the resulting weight of recovered curds (and the higher the final quantity of obtained cheese, with or without ripening).

Unfortunately, higher moisture contents mean also lower shelf-life values and enhanced proteolytic degradation (with additional demolition of organic molecules such as residual lactose and lipids). In addition, water absorption may be lowered depending on the amount of fat matter, the quantitative and qualitative composition of nitrogen-based molecules (after decomposition of caseins), and the quantity of cations with peculiar binding properties [8, 9]. On the other side, aqueous amounts may be overestimated because moisture measures do not concern water only: actually, 'moisture' means all non-solid matters that can be removed under high-temperature treatments (example: 102 °C) after a defined temporal period [10].

The study of molecular profiles of nitrogen-based molecules, caseins above all, is extremely important when speaking of cheeses, in particular packaged products. The normal detection and quantification of these special molecules is carried out in the industry (and in official laboratories by means of direct experimental methods. The main systems are [11, 12]:

(a) **The Kjeldahl method** This system is widely recognised, applicable to all possible food products, relatively cheap, and accurate. On the other side, it measures all organic nitrogen-based molecules, while protein contents might be overestimated. Moreover, the analytical result concerns only nitrogen (in grams); the determination of correlated molecules has to be made by conversion with adequate factors (example for cheeses: 6.25)
(b) **The Dumas approach** This method is very rapid and it can be performed by means of automated systems. On the other side, it can be expensive enough. In addition, it measures all organic nitrogen-based molecules, while protein contents might be overestimated
(c) **Infrared spectroscopic systems** There are different approaches depending on the used infrared spectroscopic range. Generally, these procedures are expensive enough and related equipment has to be calibrated.

On the other hand, the proteolysis of caseins in cheeses may be carried out by means of indirect systems. One of these has been discussed in recent years for a peculiar typology of cheeses: the 'Cheesemaking Yield and Proteins Estimation according to Parisi: 2006' (CYPEP:2006) method [3]. This system aims to give a simulated composition of the sampled product on the basis of two parameters only: moisture (MC) and fat matter (FC).

The aim of this Chapter is to give a description of the proteolysis evolution in selected cheeses by means of this system. A dedicated study has been carried out within a cheesemaking industry of this purpose. However, because of the important influence of water (moisture) on obtained results, this Chapter discussed only the proteolysis evolution in high-moisture cheeses only: the chosen category is mozzarella cheese obtained by cow milk curd. Other cheeses can be analysed in the same way: Chaps. 3 and 4 are dedicated to low-moisture and diced mozzarella cheeses respectively. It should be considered that diced mozzarella cheeses are normally low-moisture products; however, their moisture content appears to increase during time. Consequently, their behaviour is peculiar.

Anyway, fat matter and proteins are modified because of a notable number of chemical and biochemical reactions, in accordance with the First Law of Food Degradation [13].

2.2 Protein Evolution in High-Moisture Mozzarella Cheeses Under Refrigerated Conditions

2.2.1 Materials

Nine different productions (lots) of industrial high-moisture mozzarella cheeses have been sample for this study near a single Producer. These lots have been stored under refrigerated conditions (temperature: 2 ± 2 °C; four 400.0 grams- samples per lot) and subsequently re-sampled after seven and 14 days of storage. Consequently, six samples have been obtained for each mozzarella production (total: 54 samples). All mozzarella cheeses have been found to be vacuum-packaged with thermosealable films, generally polyamide/polyethylene plastic matters.

With reference to this study, one single high-moisture mozzarella (HMM) sample per lot has been named 'HMM-nnn' ('nnn' is an acronym used for the representation of the lot) and immediately analysed after 24 h (two samples per lot, two analyses). The remaining four samples have been named 'HMM-nnn-x' ('x' means seven or 14 days after the production) and analysed after seven and 14 days: consequently, two of them have been presented for analyses after 7 days, and the remaining cheeses have been analysed after 14 days. As an example, the third sample group for mozzarella cheeses, lot 010, analysed after 7 days, has been named 'HMM-010-7'.

All sampled cheeses have been analysed according to the above-mentioned schedule. MC, FC and proteins (PA) have been evaluated for all sampled products.

2.2.2 Analytical Methods

MC and FC have been obtained with:

(a) Infrared thermo-gravimetric method, for MC determination;
(b) AFNOR NF V04-287, for fat matter determination.

The most probable amount of proteins (PA) has been indirectly calculated by means of the CYPEP:2006 indirect method [3, 9]. All results have been calculated as the average of two data per sample.

2.2.3 *Results and Discussion*

Table 2.1 shows obtained average data (MC, FC and PA) for sampled cheeses in function of storage days after the production (1, 7 and 14 days). Obtained data correspond to the average value of the whole group of samples for MC, FC, pH and PC respectively.

Substantially, the initial composition of sampled cheeses is variegated enough, although MC is always in the range 58.4–60.6% 24 h after production. In detail (Table 2.1):

(1) MC average value is 59.5%
(2) FC value is 19.0%
(3) PA average result, obtained by means of CYPEP:2006, rigorous method, is 17.8%.

After 7 days, analytical data show a little MC augment while FC appear to be essentially unchanged; PA values seem to decrease (Table 2.1). In particular:

(a) MC increases from 59.5 to 60.8% after 7 days
(b) FC appear to remain constant: 18.7%
(c) PA values seem to tend to the number 14.7% (original result: 18.1%).

Finally, analytical data show a different situation after 14 days (Table 2.1):

(1) MC is 62.1%
(2) FC value appears to be slightly decreased (average result: 17.4%)
(3) PA indirect result is obtained as the average data of MC and FC values, in the same way of above-mentioned numbers. Interestingly, it appears unchanged: 14.1%.

Table 2.1 Chemical data for HMM samples, average data

Stored high-moisture mozzarella cheeses, refrigerated conditions (average data)			
Storage days	MC, %	FC, %	PA, %
1	59.5	19.0	17.8
7	60.8	18.7	14.7
14	62.1	17.4	14.1

MC is for: moisture, FC is for: fat matter, PA represents proteins (the most reliable amount for proteins, according to CYPEP:2006)

In general, MC profiles have shown a certain increase: from 59.5 (one day of storage) to 62.1% (14 days at 2 ± 2 °C). On the other hand:

(a) FC decreases from 19.0 to 17.4%
(b) PA values decrease in average (from 18.1 to 14.1%).

Consequently, the most probable amount of proteins in sampled cheeses shows a certain diminution (−0.9%) during 14 days; however, this decrease could have been more remarkable if compared with MC amounts (+2.6%). The augment of water in all samples, measurable as moisture content, can be easily correlated with two physical–chemical phenomena at least [5, 9]:

(a) The more or less enhanced lipolytic degradation of triglycerides in cheeses, and/or
(b) The proteolytic degradation of caseins (and the little amount of residual albumins and globulins) in the product.

The result of above-mentioned reactions may give complex situations, depending on pH, acidity values, storage conditions (high temperatures, sunlight exposure, etc.), and the involved proteolytic microflora.

The situation of packaged mozzarella cheeses (and other packaged cheeses unable to expel hydrolysis water from external cheese layers) is complex enough. The analytical composition obtained by means of CYPEP:2006 shows a tripartite emulated structure where the sum of MC, FC and PA should not exceed 100% (actually, the sum should be lower than 100 g per 100 g of sampled product).

On the contrary, examined cheeses show the following results as sum of the tripartite structure in function of storage days (Table 2.2):

(a) MC + FC + PA after 24 h: 96.3%
(b) MC + FC + PA after 7 days: 94.2%
(c) MC + FC + PA after 14 days: 93.6%.

Table 2.2 High-absorption casein content, amended high-absorption casein amount, apparent hydric absorption, the sum of the main three cheese components, and degraded protein values for HMM samples, average data

	Stored high-moisture mozzarella cheeses, refrigerated conditions (average data)				
	MC + FC + PA, %	HAC, %	A%	DP	HAC_{CORR}, %
Storage days					
1	96.3	23.9	134.0	6.1	11.7
7	94.2	24.7	168.2	10.0	4.7
14	93.6	25.0	177.4	10.9	3.2

MC is for: moisture, FC is for: fat matter, PA represents proteins (the most reliable amount for proteins, according to CYPEP:2006). HAC is high-absorption protein, A% is apparent hydric absorption, DP mean 'degraded proteins', and HAC_{CORR} means 'amended HAC'. HAC_{CORR} values have been calculated with the Italian patent-pending WISDOM Cheese software [14]

In other words, the sum of the main components in packaged mozzarella cheeses seems to decrease after 14 days with 2.7 g 'lost' per 100 g (it should be also mentioned that the labelled shelf-life of these cheeses was 28 days). Consequently, it could be inferred that the degradation of these cheeses appears be important enough.

Moreover, the chemical structure of proteins is clearly changed during time. The CYPEP:2006 indirect method can give the simulated composition of these mozzarella cheeses including 'high-absorption casein' (HAC) amounts, protein contents, and other dissolved matters (salts + residual carbohydrates + organic acids are the main constituents for this parameter).

A new quantity—apparent hydric absorption (A%)—can be calculated: this variable means the ratio between absorbed water by HAC and the aqueous content absorbed by PA. Substantially, A% represents the excessive aqueous absorption in cheeses mainly ascribed to proteins. HAC is a theoretical casein molecule with molecular weight (MW): 22,296 Da without the original glycomacropeptide fraction, which is removed in the initial stages of curd production actually representing the average weight of four different casein species in cow's milk. HAC may be apparently higher than the real PA content in cheeses; consequently, it is an apparent quantity and could be amended if needed [8, 14]. Anyway, HAC represents all caseins able to absorb water 'at the maximum level' [3, 8]. As a result, it should be considered that HAC and A% values in examined cheeses are different enough between 1 and 14 days (Table 2.2):

(a) HAC and A% after 24 h: 23.9 and 134.0% respectively
(b) HAC and A% after 7 days: 24.7 and 168.2% respectively
(c) HAC and A% after 14 days: 25.0 and 177.4% respectively.

In other words, HAC content seems to grow up during time, while the meaning of A% clearly suggest that the apparent absorption ascribed to HAC is increased if compared with the moisture amount really inglobated in the caseous matrix.

Actually, HAC can increase during time on condition that A% $\leq$ 100 only. The right approach to this problem is to consider that a remarkable fraction of hydrolysis water (and dissolved gaseous substances) cannot be expelled from packaged cheeses; caseins could mainly absorb this quantity on condition that these proteins are the only molecules able to absorb water. Naturally, this conventional assertion should be demonstrated, but water molecules in packaged cheeses can also remain 'free' (without chemical bounds such as hydrogen bonds) and consequently favour protein degradation [8].

As a result, A% may be also used as a number expressing the ratio between HAC amount in the sampled cheese (based on moisture and fat matter data) and the real PA quantity. According to CYPEP:2006 method (rigorous version), A% = 100 means that HAC = PA in packaged cheeses because:

(a) Water absorbed by HAC is equal to the aqueous content absorbed by PA, and
(b) HAC is conventionally defined able to absorb 100% of the total aqueous content until a theoretical limit equal to 3.1 kg of water per kilogram of HAC [3, 9].

Actually, A% can exceed 100% because a certain amount of water can be found in cheeses without stable hydrogen bonds with protein molecules; as a result, this 'free' water is not considered when speaking of A % $\leq$ 100 [9].

On these bases:

(1) A % values > 100 mean that a certain amount of water is not absorbed by caseins and proteins; at the same time, the difference '(A%—100)/100' may be assumed as the percentage of degraded PA proteins (DP) with MW between 0 and 22,296 Da (Eq. 2.1). On the other side, the remaining amount is equal to the amended and reliable HAC able to absorb water 'at the maximum level', also defined HAC_{CORR} (Eq. 2.2):

$$DP = PA \times \frac{(A\% - 100)}{100} \tag{2.1}$$

$$HAC_{CORR} = PA - DP \tag{2.2}$$

(2) A% values $\leq$ 100 means that water absorbed by HAC is lower than the total water absorbed by proteins. Consequently, PA are low-absorption proteins if compared with HAC: substantially, they correspond to the sum 'HAC + p-HAC' where 'para-high-absorption casein' (p-HAC) represents caseins with molecular weight > 22,296 Da. This p-HAC corresponds to the hypothetical HAC without the original glycomacropeptide fraction, which is removed in the initial stages of curd production, but it is not able to absorb notable water amounts [9]. p-HAC is present in the original cheese but is not able to absorb water 'at the maximum level'; it can only evolve towards HAC. At this stage, HAC is not proteolysed; there is no necessity of calculating HAC amended amounts.

With relation to packaged cheeses, A% apparently increases (hydrolysis water cannot be expelled and analytically eliminated) and the general trend is always the augment of A% and apparent HAC. Consequently, HAC_{CORR} and DP have to be calculated if needed (Eq. 2.1 and 2.2; Table 2.2) and possibly displayed with PA vs time (Fig. 2.1). HAC_{CORR} values have been calculated with the Italian patent-pending WISDOM Cheese software [14]. In general, DP appears to augment between 1 and 14 days (initial value: 6.1%; after 7 days, 10.0%; after 14 days, 10.9%) as the consequence of hydrolysis and apparent A% increase, while HAC_{CORR} decreases: initial value: 11.7%; after 7 days, 4.7%; after 14 days, 3.2%).

This theoretical approach has to be evaluated carefully because of the involved hypotheses: more research is needed. However, an interesting development of the simulated calculation concerns more degraded proteins, and polypeptides in particular. Scientific literature reports that one of the main amino acids found in caseins is lysine [15] with MW 146.19 Da. Substantially, a polypeptide with 15 lysine units could be hypothesised for simulation purposes and named 'simulated polypeptide'

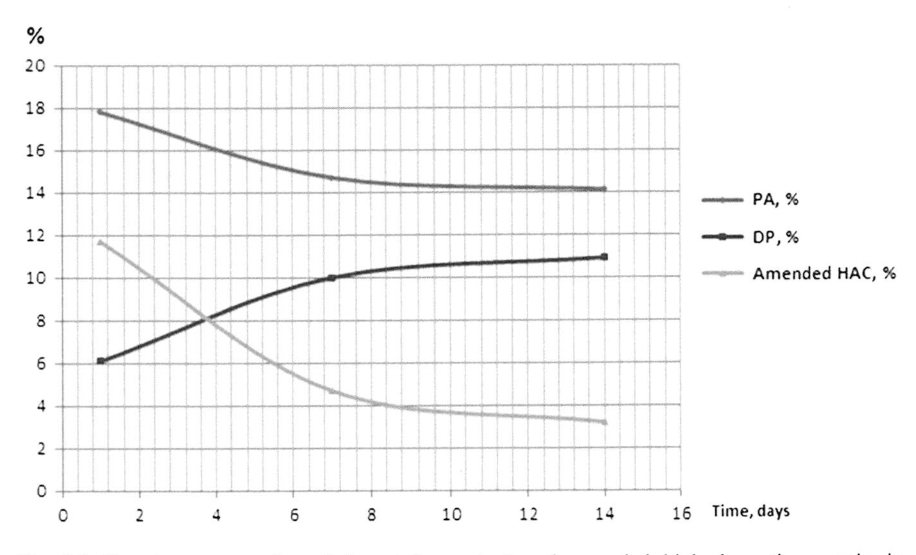

Fig. 2.1 Protein amount, degraded protein content and amended high-absorption casein in industrial high-moisture mozzarella cheeses stored under refrigerated conditions (average data) versus time (days). PA represents proteins (the most reliable amount for proteins, according to CYPEP:2006); HAC is high-absorption protein and DP mean 'degraded proteins'. Amended HAC (also named HAC_{CORR}) values have been calculated with the Italian patent-pending WISDOM Cheese software [14]

(SimP), with MW = 2192.85 Da (very close to 2223 Da). This amount might be calculated and the related evolution could be expressed versus time.

On these bases, should the DP amount be considered for each experiment and subdivided mathematically in ten different intervals with MW of 2223 Da (the HAC MW is 22,296 Da; the interval between 0 and 22,296 Da can be subdivided in ten 2223 Da- intervals), the related probable amount could be simulated on condition that:

(a) DP amount is considered as the sum of ten different quantities in the following way:

$$DP = \sum_{i=1}^{i=10} DP_i = DP_1 + DP_2 + DP_3 + \ldots + DP_{10} \tag{2.3}$$

where DP_1 is the DP amount for the 0–2223 Da- interval, DP_2 is the DP amount for the 2224–4460 Da- interval, … and DP_{10} corresponds to the DP amount for the final amount excluding HAC

(b) Each DP_i quantity is obtained as the sum of 'i' fractions (F) obtained by means of Eq. 2.4:

$$F = \frac{DP}{\sum_{i=1}^{i=10} i} = \frac{DP}{(1+2+3+4+5+6+7+8+9+10)} = \frac{DP}{55} \quad (2.4)$$

Consequently, should DP amount be equal to a, the following results could be calculated:

- F = DP/55 = a/55
- DP_1 = 1 × F = a/55
- DP_2 = 2 × F = 2a/55
- DP_3 = 3 × F = 3a/55
- …
- DP_{10} = 10 × F = 10a/55

This simulated approach can be used with the aim of displaying the theoretical degradation of PA (original HAC) during time. Figure 2.2 shows the situation for cheeses samples after 24 h, while Fig. 2.3 and 2.4 concern cheeses after seven and 14 days respectively.

Finally, the simulated low-MW-polypeptide, SimP, may be estimated (Fig. 2.2, 2.3 and 2.4): it would correspond to the first interval between 0 and 2223 Da, including all nitrogen-based organic molecules derived from proteolytic reactions in

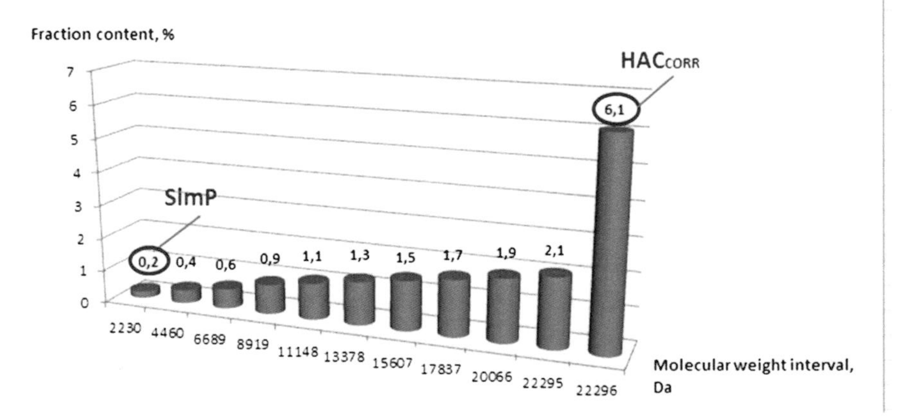

Fig. 2.2 The theoretical degradation of proteins (original HAC) in industrial high-moisture mozzarella (HMM) cheeses samples (0–4 °C) after 24 hours-storage. The lowest molecular weight interval (between 0 and 2223 Da) corresponds to the simulated low-MW-polypeptide, SimP, may be estimated and shown: it would correspond to the first interval representing all molecules in this range. Amended HAC (also named HAC_{CORR}) values have been calculated with the Italian patent-pending WISDOM Cheese software [14]

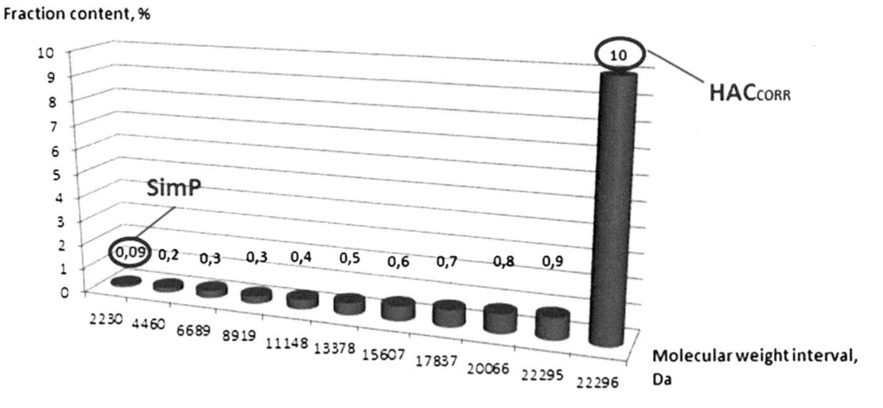

Fig. 2.3 The theoretical degradation of proteins (original HAC) in industrial high-moisture mozzarella (HMM) cheeses samples (0–4 °C) after 7 days of storage. The lowest molecular weight interval (between 0 and 2223 Da) corresponds to the simulated low-MW-polypeptide, SimP, may be estimated and shown: it would correspond to the first interval representing all molecules in this range. Amended HAC (also named HAC_{CORR}) values have been calculated with the Italian patent-pending WISDOM Cheese software [14]

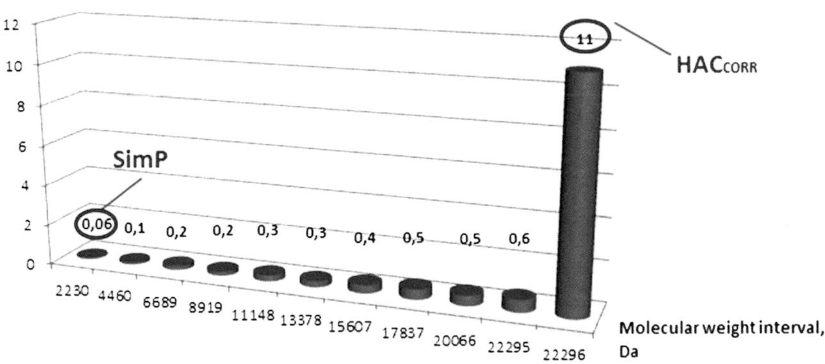

Fig. 2.4 The theoretical degradation of proteins (original HAC) in industrial high-moisture mozzarella (HMM) cheeses samples (0–4 °C) after 14 days of storage. The lowest molecular weight interval (between 0 and 2223 Da) corresponds to the simulated low-MW-polypeptide, SimP, may be estimated and shown: it would correspond to the first interval representing all molecules in this range. Amended HAC (also named HAC_{CORR}) values have been calculated with the Italian patent-pending WISDOM Cheese software [14]

this range. The emulated proteolysis is clear enough but SimP values decreased from 0.2% after 24 h to 0.09% after 7 days and 0.06% after 14 days. Substantially, the amount of small molecules increases globally during time, but demolition is mainly ascribed to medium and large protein molecules (high MW); in addition, PA decrease during time, and SimP is a protein fraction.

2.3 Conclusions

Obtained data may be interesting when speaking of high-moisture mozzarella cheeses. However, the simulation can obtain different results in other mozzarella cheeses. Chapters 3 and 4 are dedicated to low-moisture and diced mozzarella cheeses, respectively, and the same simulative approach may give different results.

References

1. Parisi S (2002) Profili evolutivi dei contenuti batterici e chimico-fisici in prodotti lattiero-caseari. Industrie Alimentari 41(412):295–306
2. Parisi S (2003) Evoluzione chimico-fisica e microbiologica nella conservazione di prodotti lattiero - caseari. Ind Aliment 42(423):249–259
3. Parisi S, Laganà P, Delia AS (2006) Il calcolo indiretto del tenore proteico nei formaggi: il metodo CYPEP. Ind Aliment 45(462):997–1010
4. Belitz HD, Grosch W, Schieberle P (2009) Food chemistry, fourth revised and extended edition. Springer, Berlin
5. Parisi S (2006) Profili chimici delle caseine presamiche alimentari. Ind Aliment 45(457):377–383
6. Delgado AM, Parisi S, Almeida MDV (2016) Milk and dairy products. Delgado AM, almeida MDV, Parisi S, chemistry of the mediterranean diet. Springer International Publishing, Cham, pp 139–176
7. Singh TK, Drake MA, Cadwallader KR (2003) Flavor of cheddar cheese: a chemical and sensory perspective. Compr Rev Food Sci Food Saf 2(4):166–189. doi:10.1111/j.1541-4337.2003.tb00021.x
8. Parisi S, Laganà P, Delia AS (2007) Lo studio dei profili proteici durante la maturazione dei formaggi tramite il metodo CYPEP. Ind Aliment 468(46):404–417
9. Parisi S (2012) La mutua ripartizione tra lipidi e caseine nei formaggi. Un approccio simulato. Ind. Aliment 523(51):7–15
10. Parisi S, Laganà P, Delia AS (2007) Microbial evolution in vacuum- packed mixtures of grated processed cheeses during cold storage. Italian J Food Sci 19:333–336
11. Alais C (1984) Science du lait. Principles des techniques laitières, 4th edn. S.E.P.A.I.C., Paris
12. Chang SKC (2003) Protein analysis. In: Nielsen SS (ed) Food Analysys, 3rd edn. Kluwer Academic and Plenum Publishers, New York
13. Parisi S, Delia S, Laganà P (2004) Il calcolo della data di scadenza degli alimenti: la funzione Shelf Life e la propagazione degli errori sperimentali. Ind Aliment 43(438):735–749
14. Parisi S, Barone C, Laganà P (2017) Procedimento per determinare la struttura di un formaggio. Brevetto per invenzione industriale (BIT), domanda numero: 102017000041086, 13 Apr 2017
15. Vickery HB, White A (1933) The Basic Amino Acids of Casein. J Biol Chem 103:413–415

Chapter 3
Evolutive Profiles of Caseins and Degraded Proteins in Industrial Low-Moisture Mozzarella Cheeses. A Simulative Approach

Abstract Mozzarella cheeses are high-perishable products if compared with other cheese typologies. One of the main reasons for lower shelf-life values in this ambit is related to protein modifications. Basic features of modern and historically relevant cheeses depend on the chemical and physical state of main components; the modification of proteins should be taken into account. With specific reference to caseins, the main nitrogen-based structure of the final cheese product, many factors influence protein degradation. Because of the needed bioavailability of free water, the amount of moisture becomes important enough. The aim of this Chapter has been to show analytical results of an emulation study carried out on different industrial low-moisture mozzarella cheeses during storage. Obtained data and the comparison with high-moisture mozzarella cheeses (Chap. 2) appear to suggest that emulated low-speed proteolysis is not important enough in low-moisture products.

Keywords Casein · Cow's milk · CYPEP:2006 · Refrigerated storage · Low-moisture mozzarella cheese · Moisture · Emulation

Abbreviations

A%	Apparent hydric absorption
CYPEP:2006	Cheesemaking Yield and Proteins Estimation according to Parisi:2006
DP	Degraded protein
FC	Fat matter
HAC	High-absorption casein
HAC_{CORR}	High-absorption residual casein
HMM	High-moisture mozzarella
LMM	Low-moisture mozzarella
MC	Moisture
MW	Molecular weight
PA	Protein
SimP	Simulated polypeptide

© The Author(s) 2018
C. Barone et al., *Chemical Evolution of Nitrogen-Compounds in Mozzarella Cheeses*, SpringerBriefs in Molecular Science,
DOI 10.1007/978-3-319-65739-4_3

3.1 Degraded Cheese Proteins and Caseins. The Role of Moisture

Cheese is substantially a tripartite matrix composed of water (determined as 'moisture content', lipids (also named 'fat matter'), and proteins, with a residual presence of carbohydrates and salt (Sect. 2.1) [1, 2]. Water is the reason which cheeses may be considered solid solutions because of abundant water contents. With relation to positive features of food products, high moisture contents may mean low durability and enhanced microbial activity with consequent degradation: lipids and proteins are mainly attacked and destroyed).

With relation to proteins in cheeses—caseins above all (Sect. 2.1) related degradation phenomena may appear more interesting because of direct and indirect influences on cheesemaking yields and safety profiles (Sect. 2.1). On the other side, similar degradations are expected during time, in accordance with the Parisi's First Law of Food Degradation [3].

The study of molecular profiles of nitrogen-based molecules, caseins above all, in cheeses may be performed by means of direct and indirect methods, including the 'Cheesemaking Yield and Proteins Estimation according to Parisi: 2006' (CYPEP:2006) method [4]. This system aims to give a simulated composition of the sampled product on the basis of two parameters only moisture (MC) and fat matter (FC).

The aim of this Chapter is to give a description of the proteolysis evolution in low-moisture mozzarella cheeses by means of this system. A dedicated study has been carried out within a cheesemaking industry of this purpose with relation to this typology and two other products: high-moisture and diced mozzarella cheeses (related results are discussed in Chaps. 2 and 4 respectively). Because of the important influence of water (moisture) on obtained results, this Chapter discussed only the proteolysis evolution in low-moisture cheeses only: the chosen category is mozzarella cheese obtained by cow milk curd [4].

3.2 Protein Evolution in Low-Moisture Mozzarella Cheeses Under Refrigerated Conditions

3.2.1 Materials

Nine different productions (lots) of industrial low-moisture mozzarella cheeses have been sampled for this study near a single Producer. These lots have been stored under refrigerated conditions (temperature: 2 ± 2°C; four 400.0 g- samples per lot) and subsequently re-sampled after 7 and 14 days of storage. As a consequence, six samples have been obtained for each mozzarella production (total: 54 samples). All mozzarella cheeses have been found to be vacuum packaged with thermosealable films, generally polyamide/polyethylene plastic matters.

With reference to this study, one single high-moisture mozzarella (LMM) sample per lot has been named 'LMM-nnn' ('nnn' is an acronym used for the representation of the lot) and immediately analysed after 24 h (two samples per lot, two analyses). The remaining four samples have been named 'LMM-nnn-x' ('x' means 7 or 14 days after the production) and analysed after 7 and 14 days: consequently, two of them have been presented for analyses after 7 days, and the remaining cheeses have been analysed after 14 days. As an example, the third sample group for mozzarella cheeses, lot 010, analysed after 7 days, has been named 'LMM-010-7'.

All sampled cheeses have been analysed according to the above-mentioned schedule (this procedure has also be explained in Chap. 2). MC, FC, and proteins (PA) have been evaluated for all sampled products.

3.2.2 *Analytical Methods*

MC and FC have been obtained with:

(a) Infrared thermo-gravimetric method, for MC determination;
(b) AFNOR NF V04-287, for fat matter determination.

The most probable amount of proteins (PA) has been indirectly calculated by means of the CYPEP:2006 indirect method [4, 5]. All results have been obtained as the average of two data per sample.

3.2.3 *Results and Discussion*

Table 3.1 shows obtained average data (MC, FC and PA) for sampled cheeses in function of storage days after the production (1, 7, and 14 days). Obtained data correspond to the average value of the whole group of samples for MC, FC, pH and PC respectively.

Substantially, the initial composition of sampled cheeses is variegated enough, although MC is always in the range 54.2–56.6% 24 h after production. In detail (Table 3.1):

1. MC average value is 55.4%
2. FC value is 18.3%

Table 3.1 Chemical data for LMM samples, average data. MC is for: moisture, FC is for: fat matter, PA represents proteins (the most reliable amount for proteins, according to CYPEP:2006)

Stored high-moisture mozzarella cheeses, refrigerated conditions (average data)			
Storage days	MC, %	FC, %	PA, %
1	55.4	18.3	21.9
7	56.2	18.1	21.6
14	57.0	17.7	21.4

3. PA average result, obtained by means of CYPEP:2006, rigorous method, is 21.9%.

After 7 days, analytical data show a little MC augment while FC appear to be essentially unchanged; PA values seem to decrease slightly (Table 3.1). In particular:

a) MC increases from 55.4 to 56.2% after 7 days
b) FC appear to remain constant: 18.1%
a) PA values seem to tend to the number 21.6% (original result: 21.9%).

Finally, analytical data show a different situation after 14 days (Table 3.1):

1. MC is 57.0%
2. FC value appears to be slightly decreased (average result: 17.7%)
3. PA indirect result is obtained as the average data of MC and FC values, in the same way of above-mentioned numbers. Interestingly, it is 21.4%.

In general, MC profiles have shown a certain increase: from 55.4 (one day of storage) to 57.0% (14 days at 2 ± 2°C). On the other hand:

a) FC decreases from 18.3 to 17.7%
b) PA values decrease in average (from 21.9 to 21.4%).

Consequently, the most probable amount of proteins in sampled LMM shows a little diminution (−0.5%) during 14 days.

The analytical composition obtained by means of CYPEP:2006 shows a tripartite emulated structure where the sum of MC, FC and PA should not exceed 100% (actually, the sum should be lower than 100 g per 100 g of sampled product). On the contrary, examined cheeses show the following results as sum of the tripartite structure in function of storage days (Table 3.2):

(a) MC + FC + PA after 24 h: 95.6%
(b) MC + FC + PA after 7 days: 95.9%
(c) MC + FC + PA after 14 days: 96.1%.

Table 3.2 High-absorption casein content, amended high-absorption casein amount, apparent hydric absorption, the sum of the main three cheese components, and degraded protein values for LMM samples, average data

	Stored high-moisture mozzarella cheeses, refrigerated conditions (average data)				
	MC + FC + PA, %	HAC, %	A%	DP	HAC_{CORR}, %
Storage days					
1	95.6	21.1	96.0	–	–
7	95.9	21.4	99.0	–	–
14	96.1	21.7	101.2	0.3	21.1

MC is for: moisture, FC is for: fat matter, PA represents proteins (the most reliable amount for proteins, according to CYPEP:2006). HAC is high-absorption protein, A% is apparent hydric absorption, DP mean 'degraded proteins', and HAC_{CORR} means 'amended HAC'. HAC_{CORR} values have been calculated with the Italian patent-pending WISDOM Cheese software [6]

In other words, the sum of the main components in packaged LMM seems to increase after 14 days (augment: + 0.5%). On the other side, HMM cheeses show a different behaviour (Sect. 2.2.3).

The CYPEP:2006 indirect method can give the simulated composition of these mozzarella cheeses including 'high-absorption casein' (HAC) amounts, protein contents, and other dissolved matters (salts + residual carbohydrates + organic acids are the main constituents for this parameter).

A new quantity—apparent hydric absorption (A%)—can be calculated (Sect. 2.2.3). With relation to HAC and A% values, the situation in LMM samples is shown in Table 3.2):

(a) HAC and A% after 24 h: 21.1 and 96.0% respectively
(b) HAC and A% after 7 days: 21.4 and 99.0% respectively
(c) HAC and A% after 14 days: 21.7 and 101.2% respectively.

In other words, HAC content seems to grow up during time, while the meaning of A% clearly suggest that the apparent absorption ascribed to HAC is increased if compared with the moisture amount really inglobated in the caseous matrix. This situation seems similar to the behaviour of high-moisture mozzarella (HMM) cheeses. Based on the approach suggested in Sect. 2.2.3, two different situations occur when speaking of LMM cheeses (Table 3.2):

(a) A% is constantly < 100 but increases between 1 and 7 days. HAC is 21.1% after 24 h and 21.4 after 7 days
(b) A% is > 100 after 14 days; as a result, DP is 0.3% and HAC has to be re-calculated as HAC_{CORR} (Eqs. 2.1 and 2.2 have been used, Sect. 2.2.3). In our situation, HAC_{CORR} is 21.1.

Consequently, HAC_{CORR} and DP have been calculated if A% > 100 (Table 3.2). In general, DP value increases only between seven and 14 days as the consequence of hydrolysis and apparent A% increase, while HAC or HAC_{CORR} values decrease.

After 14 days, the degradation of proteins could be also estimated by means of the 'simulated polypeptide' (SimP, Sect. 2.2.3). With relation to LMM cheeses, SimP is negligible: < 0.1% after 14 days (Fig. 3.1). In other terms, the emulated proteolysis is not important enough in these conditions.

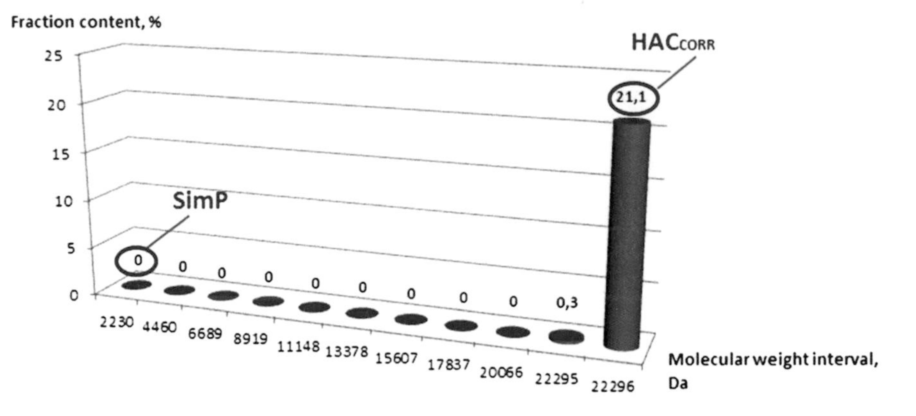

Fig. 3.1 The theoretical degradation of proteins (original HAC) in industrial high-moisture mozzarella (HMM) cheese samples (0–4°C) after 14 days of storage. The lowest molecular weight interval (between 0 and 2223 Da) corresponds to the simulated low-MW-polypeptide, SimP, could be estimated: it would correspond to the first interval representing all molecules in this range. Amended HAC (also named HAC_{CORR}) values have been calculated with the Italian patent-pending WISDOM Cheese software [6]

3.3 Conclusions

LMM cheeses show a low-speed proteolysis if compared with HMM samples, as shown by SimP, HAC and DP data. The comparison with results related to diced mozzarella cheeses (Chap. 4) might give different results.

References

1. Parisi S (2006) Profili chimici delle caseine presamiche alimentari. Ind Aliment 45(457): 377–383
2. Delgado AM, Parisi S, Almeda MDV (2016) Milk and dairy products. In: Delgado AM, Almeida MDV, Parisi S (eds) Chemistry of the Mediterranean diet, Springer International Publishing, pp 139–176
3. Parisi S, Delia S, Laganà P (2004) Il calcolo della data di scadenza degli alimenti: la funzione Shelf Life e la propagazione degli errori sperimentali. Ind Aliment 43(438):735–749
4. Parisi S, Laganà P, Delia AS (2007) Lo studio dei profili proteici durante la maturazione dei formaggi tramite il metodo CYPEP. Ind Aliment 468(46):404–417
5. Parisi S (2012) La mutua ripartizione tra lipidi e caseine nei formaggi. Un approccio simulato. Ind. Aliment 523(51):7–15
6. Parisi S, Barone C, Laganà P (2017) Procedimento per determinare la struttura di un formaggio. Brevetto per invenzione industriale (BIT), domanda numero: 102017000041086, 13 Apr 2017

Chapter 4
Evolutive Profiles of Caseins and Degraded Proteins in Industrial Diced Mozzarella Cheeses. A Simulative Approach

Abstract Cheese is substantially a tripartite matrix composed of water, fat matter and proteins, with a residual presence of carbohydrates and salt. Water is the reason which cheeses may be considered solid solutions because of abundant water contents. With relation to positive features of food products, high moisture contents may mean low durability and enhanced microbial activity with consequent degradation. Mechanical processing and related operations may be detrimental when speaking of food integrity, texture and positive properties: surely, operations such as cutting can diminish food durability, in accordance with Parisi's Second Law of Food Degradation. The aim of this Chapter has been to show analytical results of an emulation study carried out on different industrial sliced mozzarella cheeses during storage. Obtained data and comparisons with other high- and low-moisture mozzarella cheeses have shown that sliced cheeses show a very low-speed proteolysis. On the other hand, comparisons between perishable cheeses and these diced products can be questionable because of important differences such as dissimilar cheese surfaces (higher superficial areas in sliced products if compared with 'entire' cheeses).

Keywords Casein · CYPEP:2006 · Emulation · Modified atmosphere packaging · Moisture · Refrigerated storage · Sliced mozzarella cheese

Abbreviations

A%	Apparent hydric absorption
CYPEP:2006	Cheesemaking Yield and Proteins Estimation according to Parisi:2006
DP	Degraded protein
FC	Fat matter
HAC	High-absorption casein
HAC_{CORR}	High-absorption residual casein
HMM	High-moisture mozzarella
LMM	Low-moisture mozzarella
MC	Moisture
MW	Molecular weight

© The Author(s) 2018

C. Barone et al., *Chemical Evolution of Nitrogen-Compounds in Mozzarella Cheeses*, SpringerBriefs in Molecular Science,
DOI 10.1007/978-3-319-65739-4_4

PA Protein
SimP Simulated polypeptide

4.1 Degraded Cheese Proteins and Caseins. The Role of Food Processing

Cheese is substantially a tripartite matrix composed of water (determined as 'moisture content', lipids (also named 'fat matter') and proteins, with a residual presence of carbohydrates and salt (Sect. 2.1) [1, 2]. Water is the reason which cheeses may be considered solid solutions because of abundant water contents. With relation to positive features of food products, high moisture contents may mean low durability and enhanced microbial activity with consequent degradation: lipids and proteins are mainly attacked and destroyed).

Mechanical processing and related operations may be detrimental when speaking of food integrity, texture, and positive properties. Sometimes, they can be desirable. However, mechanical processes such as cutting can diminish food durability, in accordance with Parisi's Second Law of Food Degradation [3].

For this reason, diced cheeses such as sliced mozzarella (SM) cheeses can show lower shelf-life values even in peculiar conditions such as modified atmosphere packaging. Anyway, the degradation of proteins may be studied in these cheeses also by means of direct and indirect methods, including the 'Cheesemaking Yield and Proteins Estimation according to Parisi: 2006' (CYPEP:2006) method [4]. This system aims to give a simulated composition of the sampled product on the basis of two parameters only: moisture (MC) and fat matter (FC).

The aim of this chapter is to give a description of the proteolysis evolution in SM cheeses by means of this system. A dedicated study has been carried out within a cheesemaking industry of this purpose with relation to this typology and two other products: high-moisture and low-moisture mozzarella (HMM and LMM) cheeses (related results are discussed in Chaps. 2 and 3, respectively). Because of the important influence of water (moisture) on obtained results, this Chapter discussed only the proteolysis evolution in SM cheeses only: the chosen category is mozzarella cheese obtained by cow milk curd [4, 5].

4.2 Protein Evolution in Sliced Mozzarella Cheeses Under Refrigerated Conditions

4.2.1 *Materials*

Five different productions (lots) of industrial SM cheeses have been sampled for this study near a single producer. These lots have been stored under refrigerated conditions (temperature: 2 ± 2 °C; four 2500 g- samples per lot) and subsequently re-sampled after seven and 14 days of storage. Therefore, six samples have been obtained for each mozzarella production (total: 30 samples). All mozzarella cheeses have been packaged under modified atmosphere (gas composition: carbon dioxide/nitrogen 50:50) with thermosealed films and plastic boxes.

With reference to this study, one single SM sample per lot has been named 'SM-nnn' ('nnn' is an acronym used for the representation of the lot) and immediately analysed after 24 h (two samples per lot, two analyses). The remaining four samples have been named 'SM-nnn-x' ('x' means seven or 14 days after the production) and analysed after seven and 14 days: consequently, two of them have been presented for analyses after 7 days, and the remaining cheeses have been analysed after 14 days. As an example, the third sample group for mozzarella cheeses, lot 010, analysed after 7 days, has been named 'SM-010-7'.

All sampled cheeses have been analysed according to the above-mentioned schedule (this procedure has also be explained in Chap. 2). MC, FC, and proteins (PA) have been evaluated for all sampled products.

4.2.2 *Analytical Methods*

MC and FC have been obtained with:

(a) Infrared thermogravimetric method, for MC determination;
(b) AFNOR NF V04-287, for fat matter determination.

The most probable amount of proteins (PA) has been indirectly calculated by means of the CYPEP:2006 indirect method [4, 6]. All results have been obtained as the average of two data per sample.

4.2.3 *Results and Discussion*

Table 4.1 shows obtained average data (MC, FC, and PA) for sampled cheeses in function of storage days after the production (one, seven and 14 days). Obtained data correspond to the average value of the whole group of samples for MC, FC, pH and PC, respectively.

Table 4.1 Chemical data for SM samples, average data

Stored high-moisture mozzarella cheeses, refrigerated conditions (average data)			
Storage days	MC (%)	FC (%)	PA (%)
1	45.3	23.5	26.3
7	46.2	22.5	26.3
14	47.2	21.0	26.7

MC is for: moisture, FC is for: fat matter, PA represents proteins (the most reliable amount for proteins, according to CYPEP:2006)

Substantially, the initial composition of sampled cheeses is similar enough. MC is always in the range 44.0–46.6% (24 h after production). In detail (Table 4.1):

1. MC average value is 45.3%
2. FC value is 23.5%
3. PA average result, obtained by means of CYPEP:2006, rigorous method, is 26.3%.

After 7 days, analytical data show a little MC augment while FC appears to be essentially unchanged; PA values seem to decrease slightly (Table 4.1). In particular:

(a) MC increases from 45.3 to 46.2% after 7 days
(b) FC is 22.5%
(c) PA values seem to be constant: 26.3.

Finally, analytical data show a different situation after 14 days (Table 4.1):

1. MC is 47.2%
2. FC value appears to be decreased (average result: 21.0%)
3. PA indirect result is obtained as the average data of MC and FC values, in the same way of above-mentioned numbers. Interestingly, it is 26.7%.

In general, MC profiles have shown a certain increase. On the other hand:

(a) FC decreases from 23.5 to 21.0%
(b) PA values increase in average (from 26.3 to 26.7%).

Consequently, the most probable amount of proteins in sampled SM shows a little augment (+0.4%) during 14 days.

The CYPEP:2006 indirect method can give the simulated composition of SM cheeses including 'high-absorption casein' (HAC) amounts, protein contents and other dissolved matters (salts + residual carbohydrates + organic acids are the main constituents for this parameter). A new quantity—apparent hydric absorption (A%)—can be calculated (Sect. 2.2.3). Consequently, the following data have been obtained:

(a) HAC and A% after 24 h: 18.3 and 69.3%, respectively
(b) HAC and A% after 7 days: 18.4 and 69.7%, respectively
(c) HAC and A% after 14 days: 18.4 and 68.9%, respectively.

In other words, HAC and A% appear substantially constant between 1 and 14 days under refrigerated conditions. HAC is only 69–70% of the total amount of PA in cheeses; on the other hand, the 'para-high-absorption casein' (p-HAC), representing caseins with molecular weight > 22,296 Da, is constantly lower than 30% of PA. This p-HAC corresponds to the hypothetical HAC without the original glycomacropeptide fraction, which is removed in the initial stages of curd production [7]. p-HAC is present in the original cheese but is not able to absorb water 'at the maximum level'; it can only evolve towards HAC. At this stage, HAC is not proteolysed.

Consequently, SM cheese shows a certain modification of several chemical parameters, MC above all, but packaging conditions and related initial MC values do not allow observing a remarkable proteolysis. In other words, HAC does not appear degraded sufficiently (A% < 100). On the other hand, it should be noted that FC contents constantly decrease due to the probable lipolysis and emersion out of mozzarella slices [8]. As a result, this study does not appear a good test when speaking of comparison between partially proteolysed cheeses.

4.3 Conclusions

SM cheeses show a very low-speed proteolysis if compared with HMM and LMM samples, as shown by HAC and PA data. The comparison cannot be satisfactory because sliced cheeses are certainly high-surface products and consequently exposed to increased oxidative phenomena; on the other hand, modified atmosphere packaging methods act well when speaking of durability enhancement. Therefore, comparisons between perishable cheeses and these diced products can be difficult.

References

1. Parisi S (2006) Profili chimici delle caseine presamiche alimentari. Ind Aliment 45(457):377–383
2. Delgado AM, Parisi S, Almeida MDV (2016) Milk and dairy products. In: Delgado AM, Almeida MDV, Parisi S (eds) Chemistry of the Mediterranean diet, Springer International Publishing, Cham, pp 139–176
3. Parisi S, Delia S, Laganà P (2004) Il calcolo della data di scadenza degli alimenti: la funzione Shelf Life e la propagazione degli errori sperimentali. Ind Aliment 43(438):735–749
4. Parisi S, Laganà P, Delia AS (2006) Il calcolo indiretto del tenore proteico nei formaggi: il metodo CYPEP. Ind Aliment 45(462):997–1010
5. McSweeneyPLH (Ed) Cheese problems solved. Woodhead Publishing Limited, Cambridge, and CRC Press LLC, Boca Raton
6. Parisi S, Laganà P, Delia AS (2007) Lo studio dei profili proteici durante la maturazione dei formaggi tramite il metodo CYPEP. Ind Aliment 468(46):404–417
7. Alais C (1984) Science du lait. Principles des techniques laitières, 4th edn. S.E.P.A.I.C., Paris
8. Parisi S, Laganà P, Delia AS (2007) Microbial evolution in vacuum-packed mixtures of grated processed cheeses during cold storage. Ital J Food Sci 19:333–336